Joseph Jackson

**Through Glade and Mead**

A contribution to local natural history

Joseph Jackson

**Through Glade and Mead**

*A contribution to local natural history*

ISBN/EAN: 9783337212872

Printed in Europe, USA, Canada, Australia, Japan

Cover: Foto ©ninafisch / pixelio.de

More available books at **www.hansebooks.com**

# THROUGH GLADE AND MEAD

A CONTRIBUTION TO

LOCAL NATURAL HISTORY

By JOSEPH JACKSON

WORCESTER, MASSACHUSETTS
1894

# THROUGH GLADE AND MEAD

## A CONTRIBUTION TO

## LOCAL NATURAL HISTORY

By JOSEPH JACKSON

Illustrations from Photographs by J. C. Lyford

O all ye Green Things upon the earth, bless ye
the Lord: praise him, and magnify him for ever—Benedicite

WORCESTER, MASSACHUSETTS

PUBLISHED BY PUTNAM, DAVIS AND COMPANY

1894

FIVE HUNDRED AND THIRTY-FIVE COPIES PRINTED.
OF THESE THIRTY-FIVE ARE ON LARGE PAPER,
NUMBERED AND SIGNED.

QK
166
J13T

A48663
Copyright, 1894,
BY JOSEPH JACKSON.

PRIVATE PRESS OF FRANKLIN P. RICE.

Binding by
J. S. WESBY & SONS.

Half-Tones from
GILBERT G. DAVIS.

WORCESTER.

TO THE SUTTON FARMER

AND

HIS LITTLE GRANDDAUGHTER

WHO, ALTHOUGH DIFFERING MUCH IN AGE,

WERE ONE IN A COMMON LOVE

OF THE

WILD NATURE AROUND THEM IN GLADE AND MEAD

AND WHOSE LOVE IS ONLY A TYPE

OF THAT WHICH FILLS MANY HEARTS

BOTH IN CITY AND IN COUNTRY,

IN

GRATEFUL REMEMBRANCE

OF THEIR KINDLY APPRECIATION

THIS BOOK

IS DEDICATED.

# PREFACE.

It was several years ago, although it seems as if it were but yesterday, that I was walking, early in July, along the road which leads from Millbury to Sutton. In a low-lying meadow by the roadside some velvet-grass (*Holcus lanatus*, L.) was conspicuous, and I had collected a little of it for my herbarium. As I was getting over the wall again, a farmer returning from the village drove up slowly and stopped.

*Farmer.* Be you Mr. Jackson that writes them articles about wild flowers in *The Spy?*

*I.* Yes, sir.

*F.* I thought so when I see ye get over into the lot for that grass. What kind of grass is it?

*I.* (*Holding it up for his inspection.*) It is called velvet-grass in the books. I don't find anyone about here who has taken the trouble to name it in any other way.

*F.* Just so. Well, I want to tell ye, Mr. Jackson, that we take *The Weekly Spy* at my house and we're very much interested in them papers of yours.

*I.* It is pleasant to know that.

*F.* Yes, I suppose so. You see there is nobody at home but my wife and me and my little granddaughter. She's only young, but she's very much interested in the flowers all around, and when she gets hold of the paper every Saturday she just reads every word you write, and then she goes out into the fields and woods and finds all them you speak about and more too.

*I.* I am glad she is able to find so many, and that you take such an interest in them. How old is your little granddaughter?

*F.* She's only twelve, but she's bright and smart for her age. She likes to be out o' doors and learning about all these wild things. She takes solid comfort doing it. We hope you will keep right on doing as you have, because there are other folks besides us that are interested too.

I thanked him for his kindly appreciation and, after a little further conversation on outdoor matters, he drove on. I will not deny that this unexpected expression of interest and earnest injunction to "keep right on" have been sometimes in my mind in the preparation of this book, and are a sufficient explanation of the dedication.

The love of Nature is a deep-seated one. To all real children the world is a world of wonders : plant and animal life is full of mystery. To thinking adults the world is as full of wonders as it is to children, only the wonders are more wonderful, the mysteries of Life more mysterious. Contemplation of Nature is a source of endless delight to him whose eyes have in them the power of seeing. Some are born with this keen insight ; others acquire it through accidents of environment or education. In this connection I wish to express here my feeling of indebtedness to my friend of many years, Dr. George C. Webber of Millbury, in congenial association with whom and in recognizing the real gratification he drew from an all-round interest in Nature-studies I found the inspiring cause of my awaking to a deeper and more intelligent interest in these same studies.

Is it not possible that, with more rational tendencies in education, a deeper and more widely-spread interest in the lower forms of life about us may be developed? The dawn of such a day seems to be coming, to be at hand.

I take this opportunity to express my vivid sense of obligation to my long-time friend, Mr. Samuel H. Putnam of the firm of my publishers, whose persistent kindness, sympathetic encouragement and genial enthusiasm at the beginning and during the progress of this work have been a constant source of pleasure to me and will form one of my most cheering recollections.

It gives me great pleasure also to acknowledge the generous and ready help and kindly interest of my friend, Dr. Charles L. Nichols, whose love of outdoor life is not contracted by the exigencies of a busy profession.

For the artistic effects in the photographs from which the plates were prepared I am indebted to the good taste and skillful manipulation of my friend, Mr. J. C. Lyford, for whose zeal and patience in securing the best results and hearty coöperation in everything pertaining to success in this work I am truly grateful.

For the comparative freedom from errors in these pages, the typographic beauty and the artistic make-up of this book, I am indebted to the good taste and unwearying interest of my printer, Mr. Franklin P. Rice, whose unvarying courtesy and patience I wish here gratefully to acknowledge.

And now, little Book, go forth into the world and bring what pleasure to others thou canst, even as thou hast been a pleasure to me.                                                                   J. J.

WORCESTER, MASSACHUSETTS,
*November, 1894.*

# CONTENTS.

| | |
|---|---:|
| OUTDOOR LIFE IN LITERATURE | page 1 |
| THE MARCH WOODS | 20 |
| APRIL SIGNS | 25 |
| MAY FLOWERS | 35 |
| JUNE DAYS | 51 |
| THE JULY PAGEANT | 71 |
| THE AUGUST FIELDS | 89 |
| SEPTEMBER FRUITS | 104 |

## THE RECORD OF ONE YEAR.

| | |
|---|---:|
| THE EARLIEST FLOWERS | 121 |
| THE FLOWERS OF MAY | 133 |
| THE FLOWERS OF EARLY JUNE. I. | 146 |
| "        "        "        "        " II. | 160 |
| THE MID-JUNE FLOWERS | 172 |
| THE EARLY SUMMER FLOWERS | 184 |
| THE EARLY JULY FLOWERS | 195 |
| THE MID-JULY FLOWERS | 205 |

## CONTENTS.

| | |
|---|---|
| THE MIDSUMMER FLOWERS | 217 |
| THE MID-AUGUST FLOWERS | 228 |
| THE LATE SUMMER FLOWERS | 240 |

### APPENDIX A.

| | |
|---|---|
| THE FLORA OF WORCESTER COUNTY | 253 |

### APPENDIX B.

| | |
|---|---|
| THE TREES AND SHRUBS OF WORCESTER COUNTY | 321 |

## ILLUSTRATIONS.

| | |
|---|---|
| A Forest Glade | Frontispiece |
| Wild Asters | Vignette |
| Ledum latifolium | To face page 26 |
| Andromeda polifolia | " " " 40 |
| Rhododendron nudiflorum | " " " 70 |
| Swamp Meadow | " " " 114 |
| Rhododendron Rhodora | " " " 142 |
| Leucothoë racemosa | " " " 166 |
| Peat Meadow | " " " 196 |

# THROUGH GLADE AND MEAD.

## OUTDOOR LIFE IN LITERATURE.

And this our life, exempt from public haunt,
Finds tongues in trees, books in the running brooks,
Sermons in stones, and good in everything.
— *As You Like It.*

It is a familiar fact in the experience of nearly every student that the opening sentence or the first page of a new text-book is that which clings longest to the memory and rises most persistently into the realm of consciousness. The student remembers *Gallia est omnis divisa in partes tres* when the rest of Cæsar has vanished; and it is so with other authors, classical or English, whose works have been the subjects of study in childhood and youth.

In days when the study of the history of English literature was considered the proper introduction to the study of the literature itself,—Times are changed and we are changed with them,— Collier's "History of English Literature" was the text-book put into the hands of my class, and its first sentence, repeated until branded into the memory and haunting it ever afterwards, florid as it was in style, had an indefinable charm for our crude youthful taste, while it conveyed dimly to our minds the intimate connection between outdoor life and the beginnings of literature and art.

It ran as follows: "When in the depths of some Asiatic forest, shadowy with the green fans and swordblades of the palm-tribe and the giant fronds of the purple-streaked banana, a sinewy savage stood, one day long ago, etching with a thorn on some thick-fleshed leaf, torn from the luxuriant shrub-wood around him, rude images of the beasts he hunted or the arrows he shot, the first step was taken toward the making of a book." Numerous as have been the changes since those far-off days our language has preserved the fact that the tree, the beech tree probably, is the parent of the book, the leaf of the tree is the ancestor of the leaf of the book, the papyrus of the Nile Valley has given us the name for paper, and the Latin word for feather, *penna*, has given us our word for pen.

The literature of every people is redolent with the odor of outdoor life, as the forms and means by which it has been preserved speak of its outdoor origin. In the early development of our race we see the effect of the appeal of the outer world to the spiritual nature of man in the origin of the various forms of mythology. Every original expression of the devout feelings of mankind is a worship of Nature. On the plains of India, in the Nile Valley, in Syria, Greece, Italy, in Scandinavia, in that strange volcanic isle, Iceland, among the cañons of the Colorado River, and along the great plains of the West, under the Equator or the Arctic Circle, it has always been the same. The imagination, stimulated by close contact with those mighty and unknown forces of Nature, has endeavored to answer the "obstinate questionings of sense and outward things," and has peopled the earth and the air and the sea with myriad forms, beneficent or cruel. The worship of Isis and Osiris and the sacred animals has long since passed away from the Nile Valley, but the Parsee of the Iranian plateau and of western India still reverences, as his ancestors did ages ago, the glorious sun and its mystic symbol, fire. And of all forms of worship not divine, this will seem the most rational. What fitter emblem of the Almighty than the sun, on whose beneficent supply of light and heat our life depends! The men of that elder day perceived this and fell down in adoration. So every race developed its own worship, colored by the influence of the differing aspects of Nature, here bright, there gloomy, borrowing ofttimes from its neighbors, sometimes imposing its own by relentless wars.

This religion of Nature is seen at its best as developed by the highly spiritual Greeks under the clear skies of their sunny land, in the worship of the cloud-compelling Zeus, the lightning Hephæstus, the wind-ruling Æolus, the wise Athena, and the myriad other impersonations which, by a blending of physics and poetry, evolved a mythology the richness and beauty of which will be an inexhaustible source of delight through all time. The outdoor life of those ancient Greeks has reacted and is still reacting upon our own literature with a power we scarcely realize.

The old Norse myths concerning Thor and Balder and Freya and many another show us the forms under which our ancestors, dwellers by the North Sea twenty centuries ago and more, worshipped the powers of Nature. This very day was once Thor's day, dedicated to his worship; yesterday was Woden's day; to-morrow will be Freya's day; and these names, slightly modified by lapse of years, are historic landmarks of great significance in the religious development of the Germanic stock. On English soil druids once performed their religious rites in the depths of the oak groves; now Christian priests in noble cathedrals, the pillars of which are imitations of the oak boles, and the naves are imitations of the forest aisles, worship in spirit and in truth.

But the condition in which the myth is developed can exist in only a certain stage, generally an early one, in the culture of mankind. "As the Isis veil is lifted by the thinker, the gods disappear from the earth and then from the sky, and law takes their place; and yet, somehow, the sense of beauty still animates the whole and the soul feels that the divine is still near." Contemplation of Nature becomes again the worship of God, but only after all the ungodly and human has been separated from it, and naught remains but an ineffable goodness and justice and beauty.

The exigencies of an advancing civilization which tend to draw us away from the close contact with, and contemplation of, Nature rob us of a part, and that not a small one, of our lawful inheritance. It is Wordsworth, the Christian poet, who, observing this tendency, strongly expresses his feelings in a well-known sonnet:

> "The world is too much with us; late and soon,
> Getting and spending, we lay waste our powers.
> Little we see in Nature that is ours;
> We've given our hearts away, a sordid boon!
> This sea that bares her bosom to the Moon;
> The winds that will be howling at all hours,
> And are up-gathered now like sleeping flowers,
> For this, for everything, we're out of tune;
> It moves us not.—Great God! I'd rather be
> A Pagan suckled in a creed outworn,
> So might I, standing on this pleasant lea,
> Have glimpses that would make me less forlorn;
> Have sight of Proteus rising from the sea,
> Or hear old Triton blow his wreathéd horn."

This appeal of Nature to the spirit of man is, I take it, one of the most mighty of the influences of outdoor life.

In all ages there has been a marked tendency among men to gather into communities, larger or smaller, for the advantages of mutual help, protection, the gaining of wealth by mechanic arts and by trade, and the gratification of the social instincts. The current has set cityward. In prehistoric times it was as true as it is to-day, witness the ruins of great cities which have left no name behind. The glamour of city life has always existed and will exist. It may be interesting to note in this connection the fact that in the second chapter of the Bible we have a description of the garden, the Earthly Paradise, which God planted in Eden, where grew "every tree that is pleasant to the sight, and good for food"; and then after cycles of wandering and struggle, in the next to the last chapter Saint John the Divine gives us his vision of that great city, the Heavenly Paradise, the holy Jerusalem, whose "light was like unto a stone most precious, even like a jasper stone, clear as crystal, . . and the foundations of the wall of the city were garnished with all manner of precious stones"; and best of all "there shall be no more death, neither sorrow, nor crying, neither shall there be any more pain." Is not that the realization of the highest aspiration of the race, that attaining to the gates of the celestial city and entrance therein, after the long and weary struggle from the garden gate?

The fame of some great cities, of their wealth and luxury, has been one of the causes of their downfall. This was one of the charms which lured the Barbarians toward Rome. Charles Kingsley in his "The Roman and the Teuton" has finely represented this influence of the city Rome and Italian life — *Romani nominis umbra* — upon the Germanic race, the denizens of the forests beyond the Alps:

"Fancy to yourself a great Troll-garden, such as our forefathers dreamed of fifteen hundred years ago; — a fairy palace, with a fairy garden; and all around the primeval wood. Inside the Trolls dwell, cunning and wicked, watching their fairy treasures, working at their magic forges, making and making always things new and strange; and outside the forest is full of children, such children as the world had never seen before, but children still; children in frankness and purity and affectionateness and tenderness of conscience and devout awe of the unseen; and children, too, in fancy and silliness and ignorance and caprice and jealousy and quarrelsomeness and the love of excitement and adventure and the mere sport of overflowing animal health. They play unharmed among the forest beasts and conquer them in their play; but the forest is too dull and too poor for them, and they wander to the walls of the Troll-garden, and wonder what is inside. One can conceive easily for oneself what from that moment would begin to happen. Some of the more adventurous clamber in; some, too, the Trolls steal and carry off to their palace. Most never return, but here and there one escapes out again and tells how the Trolls killed all his comrades; but tells, too, of the wonders he has seen inside, of shoes of swiftness and swords of sharpness and caps of darkness; of charmed harps, charmed jewels and, above all of the charmed wine; and after all, the Trolls were very kind to him — see what fine clothes they have given him — and he struts about awhile among his companions; and then returns, and not alone. The Trolls have bewitched him, as they will bewitch more. So the fame of the Troll-garden spreads; and more and more steal in, boys and maidens, and tempt their comrades over the wall, and tell of the jewels and the dresses and the wine, the joyous maddening wine, which equals men with gods; and forget to tell how the Trolls have bought them, soul as well as body, and taught them to be vain and lustful and slavish, and tempted them, too often, to sins which have no name."

The forest-dwellers conquered in the end, but they staggered for a long time under the burdens of their victory.

All of our life is, more or less, an outdoor life. In many parts of the world, especially among the less civilized peoples, a large part of the waking hours is spent in the open air. Among the highly civilized nations there must be many persons whose daily vocations are carried on under no roof-tree. But the tendency of city life, with its special kinds of labor, must be to shut men out from the beneficent influences of air and sky. The tendency to exclude women from outdoor pursuits is still stronger. Indulgence in them can be tolerated as a recreation only, and recreation sought merely as such is apt to be a wearisome task. Homer represents Nausicaa, the daughter of Alcinous, king of Phæacia, attended by her maidens, going to the river-side to wash her bridal robes, and afterwards engaging with her companions in a game of ball, while the sage Penelope, the wife of Ulysses, attended by her maidens, is plying the loom and superintending her domestic affairs. No princess now, except some dusky African one, imitates the fair Nausicaa, nor does the modern Penelope need to weave the cloth to supply her family, but the labor of women is none the less diminished and none the less exacting. If it wears less upon the muscle, it wears more upon the nerve; and nerve is the finer and costlier fibre. It is a vital question how to repair the delicate nerve fibre, or better, how to prevent it from wearing unduly. Outdoor life, change of scene, the calm, sweet influence of earth, air and sky are the means found most effective. The gymnasium is well. It supplies an opportunity for regular, systematic exercise and training. It furnishes the skilful guide and teacher. It develops muscle, power, grace. It refreshes and strengthens. But it cannot do all. It does not make such a claim. It offers itself as a substitute. A bath-tub full of sea-water lacks something of being old ocean itself, yet it contains water enough for the purpose sought; it lacks the dash of the surf, the health-laden breeze, the bright sky overhead, "the multitudinous laughter of the ocean fields."

The problem is how to bring outdoor life near enough to us to grasp. We cannot give up the charm of the society of our kind. If we go into the country we must take the city with us, or we are in danger of realizing Alexander Selkirk's feelings as expressed by Cowper:

> "O, Solitude ! where are the charms
> That sages have seen in thy face?
> Better dwell in the midst of alarms,
> Than reign in this horrible place."

Outdoor life must call the senses into vigorous use in order to be most beneficial. At the same time it makes our senses more acute by its training. The untrained eye may be no more than a mirror, which reflects but does not observe. It sees what it brings with it the power of seeing, what it has been accustomed to see, what it is looking for. Mrs. Barbauld's familiar story, "Eyes and no Eyes, or the Art of Seeing," illustrates my meaning: Robert comes back wearied with his walk; William comes, refreshed by a delightful ramble, every step of which brought a new pleasure.

If the object sought is rest merely, it may be found in outdoor life better than elsewhere. The view of a broad New England landscape, especially on a summer day, with its meadow intervals, sparkling streams or lakes, with its distant hills or mountains woody-sloped, the various shades of green which clothe it, from the sombre colors of the pine to the bright, fresh green of the waving corn, here a broad band of brightest sunlight, there a dark shadow moved as by an invisible hand over it, the cattle browsing quietly on the hillside or standing in the cooling stream,— the view of such a landscape, lovingly gazed upon, cannot but bring peace to the wearied soul and rest to the tired body. A landscape is not a page of a book, to be taken in with a glance of the eye. It is the book itself, and the leaves must be turned if the book is to be read. It is never the same for long at a time, nor produces the same effect, although it may impress us equally with its constant beauty. Under the blush of early sunrise it presents one of its features; under the quiet of high noon, when

> "The grasshopper is silent in the grass,
> The lizard, with his shadow on the stone,
> Rests like a shadow, and the cicala sleeps,
> The purple flowers droop, the golden bee
> Is lily-cradled,"

it presents another; and when the westering sun begins to cast long shadows over it, and "The lowing herd winds slowly o'er the lea," and "The ploughman homeward plods his weary way," there is still another, all diverse, and each having its peculiar charm and producing its own effect on the mind. In summer and in winter, in autumn and in spring, we find the same diversity. It is not a cyclorama crowded with pictures at which we may look for a long time without exhausting all their meaning; it is rather the moving panorama. In music we should say it is the same sweet strain set to infinite variations.

There are two kinds of observation of Nature, one in which the results are expressed in technical language repelling the general reader, the other expressed in finished prose or poetry. The former is seen in the voluminous reports of learned societies and in government publications, exceedingly valuable in promoting the spread of knowledge; the latter is seen in the more familiar descriptive literature, which need not be a whit less scientific, although it may avoid the technicalities of the former. The difference between the two is hinted at in this little song from Tennyson's "Maud":

> "See what a lovely shell,
> Small and pure as a pearl,
> Lying close to my foot,
> Frail, but a work divine,
> Made so fairily well
> With delicate spire and whorl,
> How exquisitely minute,
> A miracle of design !
> What is it? A learned man
> Could give it a clumsy name.
> Let him name it who can,
> The beauty would be the same."

The literature of the East abounds in figurative language in which the metaphor and the simile show this observation of Nature. The unknown writer of the "Book of Job" displays such clearsightedness, such skill in description, such sincerity, such simplicity, that the "epic melody" of his work has never been surpassed. Where can be found such a picture of the war-horse, of behemoth, of the ostrich? He has learned that no man can bind the sweet influences of the Pleiades or loose the bands of Orion; he has seen visions, natural and spiritual; he has studied deeply the great problem of man's destiny and God's ways with him here on this earth, and has expressed it in free-flowing outlines.

The Great Teacher enforces his teaching by parables and gives his hearers the deepest truths, involved in the story of the sower, of the man .who planted a vineyard, of the net cast into the sea; the tree by the wayside, the lilies of the field, the birds of the air, nothing is too simple or too familiar, if only the multitude have eyes which see and ears which hear.

Homer and Theocritus and Pindar and Virgil and Ovid and other poets of classical antiquity draw their illustrations from the familiar outdoor life of their own lands, illustrations which have an equal force and meaning everywhere. Among the plants of which Homer speaks are the olive, the oak, the beech, the ash, the elm, the plane, the lotus, nepenthe, moly, asphodel, the poppy, the crocus and the hyacinth. In the Olympic games, the most famous athletic contests of all time, the victor's prize was a wreath of wild olive; in the Isthmian games, a wreath of parsley; in the Pythian games, a wreath of bay; in the Nemean games, a wreath of pine; and great honor went with them. The world was young to those old poets and they did not scorn the myth and the fairy tale, the legends of sea-gods and nymphs and enchantresses, neither mortal nor divine; and Homer, at least, loves to relate the marvels of far-off lands, tales of old sailors, it may be, who had passed far to the gates of the Læstrygones, "where such a narrow rim of night divided day from day that a man who needed not sleep might earn a double hire, and the cry of the shepherd at evening driving home his flock was heard by the shepherd going out in the morning to pasture." What fascinating pictures, too, we get of that under world where, amid "an ampler ether, a diviner air," heroes and wise men roam over the grassy meads of the Elysian fields where violet, lily and asphodel bloom forever, and rejoice their souls in sweet converse of their deeds in the upper world that had passed.

English literature abounds with charming descriptions of Nature. In that luxurious land where life has flowed for so many in quiet channels, there have been many opportunities for that loving observation so often recorded in fitting language. Such little volumes as Walton's "Complete Angler," White's "Natural History of Selborne" and Miss Mitford's "Our Village" are expressions of the English love of outdoor life. The list of English poets from whose writings might be gathered apt illustrations of this theme, from Chaucer through five centuries to Tennyson and Morris, is a long one. Just as when one who walks through forest glade or dewy mead comes unexpectedly upon some rare new orchid or fern, or sees flashing in the sunlight like a ball of flame that woodland beauty, the scarlet is this part which the reader, anxious to see how the plot ends, passes quickly and thoughtlessly over, yet not without loss to himself, for he thus neglects some of the author's best work. It is the setting to the gem, the frame to the picture, the binding to the book; it enhances its value and beauty.

Our American literature is so recent that its names are fewer, but they hold a high place. As types I will mention only Thoreau and Burroughs among our prose writers, and Bryant and Emerson among our poets. There has yet been no finer description of a naturalist as distinguished from a biologist than that given by Emerson of Thoreau:

> "And such I knew, a forest seer,
> A minstrel of the natural year,
> Foreteller of the vernal ides,
> Wise harbinger of spheres and tides,
> A lover true who knew by heart
> Each joy the mountain dales impart."

Thoreau led the way, and the number of his disciples is increasing, though they follow the master with unequal steps. Where a hundred persons read one of Thoreau's books on their publication, a thousand have now learned to look forward with pleasure to a new outdoor book by Burroughs. It is one of the signs of the times.

Emerson's "Rhodora," "Monadnoc," "The Snow-storm," "May-day" and "Wood-notes" are favorites with lovers of Nature, and so, too, are Bryant's "Thanatopsis," "The Painted Cup," "The Fringed Gentian" and "Lines to a Waterfowl." The influences of life at Concord speak in the former, and of the quiet and peace at Cummington in the latter. The high-water mark of poetry, to rise beyond which some "Storm and Stress" period of a later civilization will be needed, was reached in "Thanatopsis," a marvellous production for a youth of eighteen. American poetry may be said to begin with this poem as its spring, and the stream has rolled on with ever-widening course.

Access to the best literature is now comparatively easy. The reading of it properly may open our eyes to see for ourselves the beauties which such guides point out. Herein lies the value of a good book. It stimulates its readers to self-activity. And so, a good outdoor book may be to its readers a gateway through which they may pass into the Elysian Fields of Nature-study and Life.

# THE MARCH WOODS.

The long white lines which stretch along the brown hillsides are the relics of the snowdrifts which still lie on the shady side of the stone walls, at this distance invisible. If we could forget for a while that this is March we might think it some late October day or some long-belated bit of Indian summer arrived at its destination. The air is calm but with a chill in it suggestive of snowfields yet lingering about us. The sky is almost serene, in pleasant contrast with the most of March hitherto. This, with the cheery warbling of the bluebird and the robin, reminds us that it is time to be away to the woods again to see what signs are visible of the awakening of Nature from her winter sleep. We can hardly expect blossoms as yet, but feel sure we shall not be without some reward of our labors.

The prospect from the top of our favorite hill has much in it which reminds us of the late autumn. The nearer hills are for the most part robed in brown, with here and there patches of dark evergreen pines or spruces, while distance casts a faint purple mantle over the more remote hills which bound the horizon on every side. But it will not be long before the brown fades slowly and almost imperceptibly into the green of the later spring.

How bravely these withered stems of golden-rods by the roadside have withstood the wintry blasts! See the seeds fly now as we lightly brush past them, yielding to a slight touch what they have refused to old Boreas with all his bluster. By and by under the gentle influence of sun and moisture, they will bend to the ground, "to be resolved to earth again, to mix forever with the elements," to pass through new cycles of being. It is only a repetition of the old fable of the struggle between the sun and the wind for the traveler's cloak.

As we follow the path into the woods the snow is found lying at considerable depth, in marked contrast with the bare hill-top from which we have descended, but here and there are open spaces where we can get a glimpse of mother earth in her wild haunts once more. Our old friend, the mountain laurel (*Kalmia latifolia*, L.), looks hardly as sleek and glossy as when we bade him good-bye for the winter. But no wonder. Let us be thankful he has come out of it so well, and has in him so large a promise of beauty in the leafy month of June. We shall watch him with interest, soon putting on his new dress of the old, old fashion for the summer, and then discarding that which he now wears, so that he may be on a footing of equality with the deciduous trees, his neighbors, who are now preparing their new spring suits. How much brighter he seems than his neighbor, the white oak (*Quercus alba*, L.), who has clung to his mantle of withered brown leaves all winter!

The murmur that comes up from among the pines in the hollow tells us that the brook is certainly awake. Yes, and how it has grown! Its old home is all too narrow for it to-day. Around the boulders where it can, over them where it must, it goes chattering along; slowly widening out here, as though it had all time to loiter in; there hastily drawing itself up and plunging down to the depths below, as if it were in a hurry to overtake yesterday; but always lining its banks with a narrow fringe of green that is very pleasant to see. With what a soft carpet it has covered those rocks piled up irregularly, so complete that we can scarcely realize that there is rock beneath! As it washes around the base of that oak see how it nourishes the life not only of the tree but of those half dozen or more species of moss that, with their bright fresh green, would seem to make this a fit abode for some nymph of the stream. Through its clear waters we see scattered stems of the water cress (*Nasturtium officinale*, R. Br.) lazily waving to and fro as the busy waters hurry onward.

Soon, on this very spot where the waters have now overflowed, we shall be looking for the anemone (*Anemone nemorosa*, L.), for the star-flower (*Trientalis Americana*, Pursh), the fringed polygala (*Polygala paucifolia*, Willd), and other early flowers. If we should follow the brook down to its junction with the river, we should expect to find the skunk cabbage (*Symplocarpus fetidus*, Salisb.) already putting forth its promise of flower and leaf; but our objective point to-day is the pond yonder, by the margin of which we have often found one of the most attractive of the spring flowers, the leather-leaf (*Cassandra calyculata*, Don). Its flower-buds were already formed last summer. We have gathered twigs of it in March before now, and by placing them in water have coaxed the buds to open a month or more before those left on the parent stem. One year we found it in bloom April 22d, but the next two years on May 2d, which shows that, although it is all ready to yield to the genial influence of sun and rain, it is not in much of a hurry to open and run the risk of nipping frosts. We will gather some twigs to-day, and, as we go homeward, will not fail to pluck some of those willows which are showing their little white heads so invitingly. They have pushed back the scales which covered them carefully through the winter, and now, protected by the numerous wool-like hairs, are pushing further forward so as not to lose a minute when the time has come to develop themselves. A few twigs of the alder and the hazel will reward us for any trouble by smiling for us before the general blossoming.

Last autumn we bade the blossoms of the witch hazel (*Hamamelis Virginiana*, L.) a hearty welcome after its leaves had fallen; this spring we shall bid as hearty a welcome to the flowers of the common hazel (*Corylus Americana*, Walt.) before the leaves have appeared. Although they have some resemblance in name, they are far removed botanically from each other. They occupy almost as remote positions in the seasons; the one among the foremost to appear, the other bringing up the rear. The common hazel is lured forth by the balmy winds of early April; the witch hazel comes when the sun has returned to the south, and cold winter is threatening to fold us in his icy embrace.

The scouts of the advance guard of the annual flower procession are in sight. The main body is not far behind. We shall need to be alert, or some parts of the gay pageant will escape our eyes.

# APRIL SIGNS.

> O fair midspring, besung so oft and oft,
> How can I praise thy loveliness enow;
> Thy sun that burns not, and thy breezes soft
> That o'er the blossoms of the orchard blow,
> The thousand things that 'neath the young leaves grow,
> The hopes and chances of the growing year,
> Winter forgotten long, and summer near?
> — WILLIAM MORRIS — *The Earthly Paradise.*

The dead leaves in the glade are soft with the warm April rain and do not rustle as once they did. Their airiness, their readiness to answer to the call of the winds is gone; the burden of the winter snow has pressed upon them and crowded them down to the dark earth of which they are beginning to form a part. But beneath them life is awaking, the buds underground as well as above are swelling, myriad forms are striving to reach up into the light.

Situated as we are, about half-way between the Equator and the North Pole, in one of the most highly favored latitudes, on almost the same parallel with Constantinople, Rome, Madrid, seats of the Old World civilization's bright consummate flower, it is not strange that our flora should be a varied one, partaking of both a northern and a southern character and containing species of world-wide distribution, with tropical and polar kindred. One never realizes how varied and abundant it is until some especial opportunity or some especial interest leads him to investigate carefully. Then, as the eye sees what it is looking for, ever new and farther-reaching vistas open before it, and the distant seems to come near.

Ralph Cranfield, in Hawthorne's fairy legend of "The Three-fold Destiny," finds at the very door of his mother's dwelling the treasure he has sought in worldwide wanderings. One of my friends who has sailed toward the far North, to the Greenland coast, has often told me of the strangeness of that wild land where, close to the foot of the great blue ice-masses, the bright summer flowers bloom; he had brought home sprays of one of those humble shrubs which, without blossoms, we had concluded must be the Labrador tea (*Ledum latifolium*, Ait.). In that now classic work, the "Florula Bostoniensis," the third and best edition of which appeared in 1840, Dr. Bigelow describes the Labrador tea, but is obliged to refer to Mount Monadnock and the White Mountains as its nearest known habitats. Six

*Ledum latifolium*, Ait.
*The Labrador Tea.*

*The northlands claim with pride thy blossoms fair,
And yet thou spurnest not our summer air.*

years later Mr. George B. Emerson, in his most excellent "Report on the Trees and Shrubs growing naturally in the Forests of Massachusetts," has either seen it or heard of it growing in Pittsfield, Richmond and Hubbardston; but now I know it is equally at home in one sheltered spot within three miles of our City Hall. The first one-flowered pyrola (*Moneses grandiflora*, Salisb.) that I ever saw was sent to me by a friend spending the summer at Franconia, New Hampshire; yet it was not very long afterwards that it was found in the woods on the side of Rattlesnake Hill. I should be surprised to find a grove of coco-nut palms (*Cocos nucifera*, L.) in the neighborhood of Worcester, but I am not surprised to find a Lilliputian imitation of it in a handful of that interesting liverwort (*Marchantia polymorpha*, L.), everywhere common.

Some plants are so conspicuous and so common, have such an innate attractive quality, that they have found a place in the literature and the life of the people. They are wrought into its art, its poetry and its legends. It is the native flora, however, and not the exotic, which thus wins its way into the hearts of the people. In the poet's verse, in many different lands, are embalmed sweet and sunny memories, it may be of the date-palm among the Arabs, of the rose among the Persians, of the lotus among the Egyptians, of the olive among the Greeks, of the fleur-de-lis among the French, of the eglantine, the heather and others among the English, of the violet, the trailing arbutus, the painted-cup, the golden-rod and others among ourselves. The native flora is that which is associated with most of our recollections of Nature. Fields of buttercups and daisies belong to the memories of spring always, wild roses and wild berries to the summer, asters and golden-rods to the autumn. It is no wonder that, after that first severe winter at Plymouth, the Pilgrims gave to the early flowers and the early bird-visitors names which recalled the scenes of their childhood in that far-off old home.

The great event of the year is the sun's crossing the Equator. The vernal equinox, in a sense the initial point of the year, is surely the initial point for plant and animal life. It brings longer days; it brings sunshine, warmth, light, hope. It strikes the key-note in the great anthem of Life, which will rise in grand and ever grander strains through the few coming months. It is the herald of many beauties.

> "In the Spring a fuller crimson comes upon the
> robin's breast;
> In the Spring the wanton lapwing gets himself
> another crest;
> In the Spring a livelier iris changes on the
> burnished dove;
> In the Spring a young man's fancy lightly turns
> to thoughts of love."

Nearly all flowering plants are divided into the two great groups of wind-fertilized (*anemophilous*) and insect-fertilized (*entomophilous*) flowers. In the early spring, while the hordes of insects are not yet astir, but while the winds are never sleeping in their efforts to restore the equilibrium of the atmosphere disturbed by the northward progress of the sun, the anemophilous flowers hasten to open. The catkins of the alders, willows, poplars, hazels, birches, oaks, walnuts, butternut, beech, sweet-fern, sweet gale, bayberry, hornbeam and hop-hornbeam, which were formed during the preceding summer, now scatter their fertilizing pollen freely and abundantly to the wooing breezes. These flowers lack the color and the perfume which render so many entomophilous flowers so attractive. Not depending on the visits of insects to secure fertilization, they do not need the attractive charms which would bring them. They do need projecting stamens laden with pollen, most of which may be wasted—for Nature is in many ways reckless in the use of her material,—but enough will remain to accomplish the desired purpose. The fact that the leaves of these trees and shrubs are not yet expanded would seem to indicate that Nature does not wish to interpose unnecessarily any obstacle to the pollen on its journey to the waiting stigma.

But the general interest does not lie in such flowers as these, consisting as they do mainly of the essential organs, the stamens and the pistils, and lacking the showy calyx or corolla. It is rather in such as the trailing arbutus (*Epigæa repens*, L.), the hepatica (*Hepatica triloba*, Chaix)—I dislike very much its book-name, liverleaf—the bloodroot (*Sanguinaria Canadensis*, L.), the anemone (*Anemone nemorosa*, L.)—who thinks of calling it windflower?—the early yellow violet (*Viola rotundifolia*, Michx.). There is a delicate beauty about these which endears them to all hearts. Year after year the pale pink blooms of the trailing arbutus allure us to some favorite and well-remembered nook where the sweet and quiet eyes are opening under the last year's dead leaves. What pleasant recollections of long-gone springtimes are associated with this little flower! The tender grace of those days that are dead does come back to us and we feel the touch of the vanished hands and hear again the voices that are still. It is easier to pull up this favorite flower by the roots than to break the stems, and hence comes the necessity of forbearance. It is worth while to be able to answer in the affirmative the questions of our philosopher-poet, Emerson:

> "Hast thou named all the birds without a gun?
> Loved the wood-rose, and left it on its stalk?
>
> . . . . . .
>
> O, be my friend, and teach me to be thine!"

Not so well known nor so much sought for, but equally attractive, is the hepatica, of which the pale blue or white blossoms peering among the tri-lobed downy leaves are the prize of the searcher for woodland beauties. Of evanescent beauty, dropping its two sepals before the petals are fully expanded, and dropping its petals while you are plucking it, the white-flowered, yellow-stamened bloodroot by many a brookside is interwoven as one more thread into the living garment of the Deity. Not many of our towns can boast of more than two or three localities for finding all of these.

In the deep woods it may be that we shall find late in April one of our rare shrubs, which is much more abundant further north, the leatherwood (*Dirca palustris*, L.). Its clusters of small yellow flowers precede the leaves, so that they are rarely seen except by those who are seeking for them. The wood of this shrub is brittle, but the bark is exceedingly tough. Leatherbark would be a more appropriate common name. While trying to find it once near where it had been found before, we came upon a decaying stump which was all covered with the delicate green of one of our most graceful little mosses (*Tetraphis pellucida*, Hedw.). Close by was the early yellow violet, whose praise Bryant sang, probably at Cummington, in the interval between the composition of those two matchless poems, "Thanatopsis" and "Lines to a Waterfowl." Ever since that day the sight of one of these, wherever it may be, recalls instantly, paints as it were on vacancy, the picture of a little wooded dell down which a brook runs gently murmuring, where the earliest spring flowers love to dwell. It was in this quiet nook that the earliest butterflies would be found in the last days of April; the Antiopa butterfly (*Vanessa Antiopa*, L.), having spent the winter in some sheltered spot in a partially torpid state, is fluttering with ragged and faded wings in search of its mate. With this early start two generations of Antiopa will see the light before the frosts of another winter. There too, at the same time, we have seen the small yellow Philodice butterfly (*Colias Philodice*, Godart), the Comma butterfly (*Grapta Comma*, Harris), and the beautiful little azure-blue (*Lycæna Pseudargiolus*, Boisd. and Lec.). The old Dwinnell road, which leads to this charmed spot, was in those days our favorite path. It seemed to lead into a region where

> "All that is most beauteous is imaged there
> In happier beauty; more pellucid streams,
> An ampler ether, a diviner air,
> And fields invested with purpureal gleams."

Was there ever such a road as that? Can there ever be such another? We have our doubts.

We have found in April in small ponds, just before its season is over, one of the most interesting of the Phyllopod Crustacea, the *Branchipus vernalis*, Verrill. "These Crustacea," says Professor Packard, "are of singular beauty and interest in themselves, and their movements while swimming on their backs very graceful. Moreover, when we consider the habits of all the Phyllopods; their singular means of adaptation to great changes in their environment; the great vitality of the species; when we take into account their weak and delicate individual organization, and when we note their interesting metamorphoses and many points in their structure, we are forced to conclude that the Phyllopods are the most interesting of all the Crustacea." They live mostly in pools of fresh water liable to dry up in summer. The eggs after being fertilized and borne about for a time finally drop to the bottom of the pool. Here they must lie in the dried mud, after the water has evaporated, until the autumn rains again fill the pool, the young hatch out and a new cycle begins. Those which we gathered and kept in a large glass globe lived but two or three days and appeared to dissolve in the water, leaving scarcely a trace. They have been found in southern New England from the last of November to the first of May, but have not yet been found during the summer, being then represented by the eggs alone. This might seem at first sight to indicate a precarious existence for this little group, but it is not so. No matter how many of the eggs may be destroyed in various ways, Nature provides that there shall be enough to insure the continuance of the race until its mission is accomplished and its place is needed for something higher and better.

But *Branchipus* is only one of the myriad forms of life which may be found in the pool. Microscopic forms, algæ, desmids, diatoms, rhizopods, infusoria, are there, most of them of wonderful beauty. To watch their development on a life-slide under a compound microscope is to get a clear vision into the inner recesses of "Nature's Fairy Land."

## MAY FLOWERS.

"And what was the Trinity Flower like, my Father?" asked the boy.
"It was about the size of Herb Paris, my son," replied the hermit. "But instead of being fourfold every way, it numbered the mystic Three. Every part was threefold. The leaves were three, the petals three, the sepals three. The flower was snow-white, but on each of the three parts it was stained with crimson stripes, like white garments dyed in blood."
— Mrs. J. H. EWING — *The Trinity Flower.*

It seems a long time since Howard's Woods were cut down. They had a special interest for me because I first found in them many plants whose reappearance year by year was very welcome. On the gently sloping hillside was a grove of scattered pine trees, mingled with elm and maple, amid which a little brook slipped on its quiet way down into the valley where it was hidden under the thick shade of maple, beech, hornbeam and hemlock. By its side grew some of the earliest flowers, and conspicuous among these by its size and color was the purple trillium or birthroot (*Trillium erectum*, L.). For several years I watched them in bud and blossom as they developed slowly in the warm spring days, until I felt that I knew them all and that they waited my coming. But after a time the trees were sacrificed to the needs of their owner, and the sunlight streaming in caused the little brook to shrink, yet the purple trilliums do not cease to gladden the spring with their beauty. They bloom now without me, just as for ages rare tropical orchids have bloomed in the depths of South American and Malayan forests unseen.

By the middle of May the other two species of this genus found in this county appear almost together. They are the nodding trillium or wake-robin (*Trillium cernuum*, L.), which hides its flowers beneath the leaves by curving its stem downward, and the loveliest of the three, the painted trillium (*Trillium erythrocarpum*, Michx.), the Trinity Flower of Mrs. Ewing's charming story. The trilliums have a striking family resemblance, and are easily distinguished from the other members of their order, the Liliaceæ, by the fact that the six parts of the perianth are divided into two groups of different colors, which are naturally called sepals and petals, and that the leaves are netted-veined.

When the painted trillium is in bloom it is time to look for the unfolding broad cymes of the earliest of our viburnums, the hobble-bush (*Viburnum lantanoides*, Michx.). Its leaves are symmetrical and handsome, larger than those of any other native shrub. It is worthy of notice from the fact that its leaf-buds are naked, i. e. unprotected by scales as in our other native shrubs and trees.

One of the shrubs which attracts attention in early spring, whether we ride on highway or railway or thread on foot the forest paths, is the shad-bush or June berry (*Amelanchier Canadensis*, Torr. and Gray). Its numerous white flowers in drooping racemes show very plainly through the as yet leafless woods, and indicate the abundance of the shrub or small tree much more clearly than can be done in summer when the other shrubs and trees are in full leafage. May is the month when most of the berry-bearing shrubs bloom, the fruits of which are of greater or less commercial value. The black huckleberry (*Gaylussacia resinosa*, Torr. and Gray) is one of the most important of these and at the same time one of the most abundant. Huckleberrying in rural districts is one of the events of summer. The huckleberry pasture is a prominent feature in most New England landscapes. Wilson Flagg in his delightful "The Woods and Byways of New England" attempts to fix a distinction in the use of the words *whortleberries* and *huckleberries*, which we think will scarcely be maintained. "The wild pastures abound in summer with well-known fruits, some of jet and some of azure. We go out with a few friends and gather them with flowers for present amusement. These fruits are *Whortleberries*.

This is their poetical and their botanical name, the one that is associated with all the beautiful things that cluster in the same field. These fruits are also gathered for the market, and exposed for sale with cucumbers, new potatoes, and squashes. They are now *Huckleberries*. . . . We should say Whortleberries if we are writing an essay or a poem about them, and Huckleberries if we are going to buy a few of them in the market. The usages of the market in other matters ought to be excluded from literature. In commerce, for example, fishes are fish; in natural history fish are fishes." In this case, so far as my experience goes, the usage of the market bids fair to triumph, just as in coinage the cheaper metal tends to drive the more valuable one out of circulation. The dangleberry (*Gaylussacia frondosa*, Torr. and Gray), our only other species of this genus, is not so abundant as the preceding, and is, consequently, not so well known.

Four species of blueberry are found in our county, the dwarf (*Vaccinium Pennsylvanicum*, Lam.), the low (*V. vacillans*, Solander), the common (*V. corymbosum*, L.) with its variety, and the northern (*V. Canadense*, Kalm). The low and the common blueberries are rivals, in popular favor, of the huckleberries. These shrubs are pleasant to the eye, both when covered with bloom and when laden with fruit.

The summer traveler among the White Mountains of New Hampshire or the Adirondacks of New York can scarcely fail to notice, as he climbs the mountain paths in the moss-floored woods, mats of a delicate trailing evergreen, with small short-petioled leaves, and bearing here and there a snow-white berry. He may never see it in flower, but he recognizes the creeping snowberry (*Chiogenes serpyllifolia*, Salisb.), one of the rarer plants of this region and rare further south except in the mountain regions.

Worcester County is especially rich in species of the great order Ericaceæ. There are thirty-four so far discovered, a number surpassing that of such States as Vermont, Rhode Island, Indiana, Nebraska and Texas. A few of them are of world-wide distribution. One of these is the bearberry or mountain cranberry (*Arctostaphylos Uva-ursi*, Spreng.), a native of northern Europe, Siberia and North America. Its foliage is much admired for its glossy greenness, and its red berries furnish food in winter to those birds, as the quail and the robin, which can then find insect food only with great difficulty.

One of the most beautiful and at the same time very rare members of this family, one whose presence we hardly dared to hope for, is the water Andromeda (*Andromeda polifolia*, L.). It was for this modest and delicate plant, which is a native of the north of

Europe as well as of this country, that Linnæus selected the poetical name of the genus. The following is the account which he himself gives of it in his "Tour in Lapland." "*Andromeda polifolia* was now (June 12) in its highest beauty, decorating the marshy grounds in a most agreeable manner. The flowers are quite blood-red before they expand; but, when full grown, the corolla is of a flesh color. Scarcely any painter's art can so happily imitate the beauty of a fine female complexion; still less could any artificial color upon the face itself bear any comparison with this lovely blossom. As I contemplated it, I could not help thinking of Andromeda as described by the poets, and the more I meditated upon their descriptions, the more applicable they seemed to the little plant before me. This plant is always fixed on some turfy little hillock in the midst of the swamps, as Andromeda herself was chained to a rock in the sea which bathed her feet, as the fresh water does the roots of this plant. At length comes Perseus in the shape of summer, dries up the surrounding water, and destroys the monsters about her."

The tale has been nowhere more finely told than in Charles Kingsley's poem, "Andromeda," one of the best specimens of English hexameters, in which the grey-eyed Pallas Athené blesses the nuptials of Perseus and Andromeda:

*Andromeda polifolia*, L.
*The Water Andromeda.*

*The maid Andromeda, divinely fair,*
*Forever lives in poesy: a group*
*Of stars in northern skies keeps bright her fame ;*
*This little flower each spring recalls her name.*

"Courage I give thee; the heart of a queen, and the mind
 of Immortals;
Godlike to talk with the gods, and to look on their eyes
 unshrinking;
Fearing the sun and the stars no more, and the blue salt
 water;
Fearing us only, the lords of Olympus, friends of the
 heroes;
Chastely and wisely to govern thyself and thy house and
 thy people,
Bearing a god-like race to thy spouse, till dying I set
 thee
High for a star in the heavens, a sign and a hope to the
 seamen,
Spreading thy long white arms all night in the heights of
 the æther,
Hard by thy sire and the hero thy spouse, while near thee
 thy mother
Sits in her ivory chair, as she plaits ambrosial tresses.
All night long thou wilt shine; all day thou wilt feast on
 Olympus,
Happy, the guest of the gods, by thy husband, the god-
 begotten."

The substance of the story is also given by William Morris in his "The Earthly Paradise," in the tale entitled "The Doom of King Acrisius." The poetic legend adds to our interest in the plant; the plant adds to our interest in the poetic legend.

The plant *Andromeda* blooms here about the middle of May and with its glossy foliage is one of the ornaments of summer: in the long nights of winter the constellation *Andromeda* is one of the most prominent groups in the northern sky.

A little pool in the middle of a great pasture; groups of tall hickories or chestnuts scattered here and there; little knolls crowned with huckleberry bushes; pink azaleas covered with fragrant blossoms; mats of the floating foxtail grass (*Alopecurus geniculatus*, L., var. *aristulatus*, Torr.) lying on the edge of the pool where the water is shallow; the yellow of the buttercups (*Ranunculus bulbosus*, L.) mingling with the light purple of one of the fleabanes (*Erigeron bellidifolius*, Muhl.); these are a fitting framework to enclose the picture of *Leucothoe racemosa*, Gray.

> When sweet May days are drawing to an end,
> What vision fair is this that greets mine eyes,
> As through the wildwood path my way I wend,
> Intent to win from Nature some new prize?
>
> Thou hast few rivals, fair Leucothoe,
> In grace and loveliness among the flowers,
> Thy long racemes are full of purity
> And fragrance mild, thou charm of woodland bowers.
>
> The pool is happy in whose face all day
> Thy gentle smiles are shining tenderly;

The glorious sun is cheered upon his way
To see thy beauty, fair Leucothoe.

And while we thank the Giver for His gifts,
We'll not forget thee, fair Leucothoe;
And while our heart the song of praise uplifts,
'T will be more earnest with the thought of thee.

At no other time in the year are we more sure that two of our senses will be fully gratified, the eye with color and the ear with song, for the spring-tide colors of the foliage are now at their best. There is the grey-green of the unfolding poplar leaves which at a little distance, seen through the opening foliage of other trees, resemble blossoms; the bright green of the birches, heightened by the brown and yellow of the numerous pendulous catkins; the russet of the maples, the purple of the oaks, the dark green of the pines and spruces, blending into a beautiful picture, charming the eyes as they sweep over the wide-extended prospect of upland and meadow to the distant hills that seem to stand guard over it all. The songs of the veery and the brown thrush and the summer warbler fill the solitudes with refreshing sound. May is preëminently the month for birds as it is for flowers. We naturally associate them together. Some little circumstance binds them to each other so that the one suggests the other. Some of my groups are: the chewink (*Pipilo erythrophthal-*

*amus*, Vieillot) and the purple trillium; the ovenbird (*Seiurus auricapillus*, Sw.) and the wild comfrey (*Cynoglossum Virginicum*, L.); the scarlet tanager (*Pyranga rubra*, Vieillot) and the rhodora; the whippoorwill (*Antrostomus vociferus*, Bonaparte) and one of the pogonias (*P. verticillata*, Nutt.); the ruffed grouse (*Bonasa umbellus*, Stephens) and the Jack-in-the-pulpit (*Arisæma triphyllum*, Torr.).

A chewink always used to flit along the stone wall which bordered one side of the lane leading to the home of the purple trillium. It was while looking for the pale blue flowers of wild comfrey, near the only spot in which I ever found it, that I looked into the nest of an oven-bird, which under other circumstances would not have been noticed. The rhodora and the scarlet tanager were almost always seen for the first time each season on the same day. Nothing in the landscape could give greater pleasure than the sight of these two. I could appreciate Emerson's praise of the rhodora, and Coues' tribute to the scarlet tanager as given in his "Birds of the Colorado Valley" where he says:

"I hold this bird in particular, almost superstitious, recollection, as the very first of all the feathered tribe to stir within me those emotions that have never ceased to stimulate and gratify my love for birds. More years have passed than I care to remember since a little child was strolling through an orchard one bright morning in June, filled with mute wonder at beauties felt, but neither questioned nor understood. A shout from an older companion—'There goes a Scarlet Tanager!'—and the child was straining eager, wistful eyes after something that had flashed upon his senses for a moment as if from another world, it seemed so bright, so beautiful, so strange. 'What is a Scarlet Tanager?' mused the child, whose consciousness had flown with the wonderful apparition on wings of ecstasy; but the bees hummed on, the scent of flowers floated by, the sunbeam passed across the greensward, and there was no reply—nothing but the echo of a mute appeal to Nature, stirring the very depths with an inward thrill. That night the vision came again in dreamland, where the strangest things are truest and known the best; the child was startled by a ball of fire, and fanned to rest again by a sable wing. The wax was soft then, and the impress grew indelible. Nor would I blur it if I could—not though the flight of years has borne sad answers to reiterated questionings—not though the wings of hope are tipped with lead and brush the very earth instead of soaring in scented sunlight."

*Pogonia verticillata*, Nutt. used to grow abundantly in Heywood's Woods, on the edge of a clearing again overgrown with clumps of young chestnuts. One day while I was looking carefully for pogonia among the underbrush a strange bird rose from the ground a foot or two and flew noiselessly out of sight. It was only a few steps to the spot, and there lay two of the handsomest eggs found in New England, of a delicate creamy-white with blotches, lines and spots of different shades of light-brown and lavender. There was no nest, simply a slight hollow among the dead leaves. My curiosity was aroused to discover the name of the bird, and a visit about an hour later, when she was flushed from the nest, disclosed the fact that I had stumbled upon a very rare opportunity to watch one of our little-known birds. One of the eggs was taken to add to a small collection of birds' eggs, a license having been previously obtained from the proper authority. At the end of two weeks a little ball of down was lying where the remaining egg had been. At the end of another week it had grown into the appearance of a young bird. At all these visits the mother-bird sat quietly on the nest until I had approached within ten feet of her, and then flew noiselessly away. But at the end of another week when I went to see what was the state of affairs, she ran toward me and fluttered around me while the little one scampered away among the underbrush that was near. Then when all was safe she flew away, and I saw her no more. I looked in vain in after years near the same spot, but I have not again stumbled upon so rare a sight.

The sweet vernal-grass (*Anthoxanthum odoratum*, L.) was the first of the grasses that I succeeded in making out. It was pleasant, therefore, to read in the "Life and Letters of Charles Darwin," by his son Francis, a letter written to Sir J. D. Hooker. It is dated June 5, 1855. Darwin was then forty-six years old, had published his "Journal of Researches" made during the voyage of the Beagle, and was meditating on "The Origin of Species." The letter closes as follows: "I have just made out my first grass, hurrah! hurrah! I must confess that fortune favors the bold, for, as good luck would have it, it was the easy *Anthoxanthum odoratum;* nevertheless it is a great discovery; I never expected to make out a grass in all my life, so hurrah! It has done my stomach surprising good."

In our neighborhood we may find several other flowers which have some interest from their connection with great names in the science of botany. While he was still a student at Fairfield Academy Asa Gray's passion for plants was aroused by reading the article on botany in the Edinburgh Encyclopædia. He at once bought a copy of "Eaton's Botany" and longed for spring. As the season opened, "he sallied forth early, discovered a plant in bloom, brought it home and found its name to be *Claytonia Virginica*, L., the species *Claytonia Caroliniana*, Michx., to which the plant really belonged, not being distinguished then."

This is the spring beauty, one of the favorite flowers with those who have been fortunate enough to find it.

The chance discovery of that somewhat rare and curious moss, *Buxbaumia aphylla*, L., directed the attention of Sir William J. Hooker, the organizer of the Royal Gardens at Kew on their present high basis, towards botany and fixed the bent of his long and active life.

The blue cohosh or pappoose-root (*Caulophyllum thalictroides*, Michx.), flowering at Kew, gave Robert Brown, the foremost botanist of his time, an opportunity to make the observations recorded in one of his papers, "On some Remarkable Deviations from the usual structure of Seeds and Fruits." Its seed at an early stage of its growth bursts the pericarp or envelope, the withered remains of which are in most cases visible at the base of the seed, which thus ripens naked.

Whatever can be found anywhere of botanical interest can be found represented in some form in our local flora.

"The wild marsh-marigold shines like fire in swamps
    and hollows gray"

in English counties, and it shines just as brightly under another name, cowslip, in our own meadows and lowlands. Do insectivorous plants attract the attention of the naturalist? Nearly two-thirds of Darwin's work on "Insectivorous Plants" is devoted to the common sundew (*Drosera rotundifolia*, L.). Probably no plant has been subjected to a greater variety of experiments, with the results noted more carefully, or the deductions drawn with greater deliberation. Is it a question of the wonderful contrivances by which orchids are fertilized by insects? Our native species, thirty-two in number, several blooming in May, exhibit many of them. No one can study their structure at all thoughtfully without recognizing in it a miracle of design. Our *Habenaria Hookeri*, Torr. and *Arethusa bulbosa*, L. and *Pogonia verticillata*, Nutt. are among the most interesting of these.

Weeds, with a few exceptions, do not bloom in May. They start early enough, but feeling that they have all summer and autumn to ripen in do not hurry to bloom. It may be they realize they are safer from harm by lying low until the more useful vegetation has attained a sufficient growth to overshadow and protect them. The dandelion (*Taraxacum officinale*, Weber) has a habit of growth which would seem, in the animal kingdom, to imply a great deal of instinct or inherited memory or whatever we may choose to call it. It blooms early on a short scape or flower-stem which lengthens during blossoming. After all the flowers have been expanded long enough to ensure fertilization, the inner involucre closes and the cells on one side of the stem contract, drawing it down almost to the ground. There and then the slender beak elongates and raises up the pappus or crown of soft white hairs, and the fruit ripens. Then the cells expand again, the stem stands erect, the involucre is reflexed, and the achenes or seed-like fruits with the pappus displayed in an open globular head are ready to be carried by the lightest wind to form a colony away from the parent plant. The shepherd's purse (*Capsella Bursa-pastoris*, Moench.), and chickweed (*Stellaria media*, Smith), and field sorrel (*Rumex Acetosella*, L.) are the weeds of May, which have matters almost their own way in the early part of the season.

But we cannot even mention all the flowers of May. The violets must pass unheeded, with the Solomon's seal, and the smilacinas, and the dwarf ginseng, and the fringed polygala, and the gold thread, and the one-flowered cancer-root, and iris, and lupine, and starflower, and blue-eyed grass, and yellow star-grass, and columbine, and bellwort, and wild oats, and wild calla. Next year the procession will pass by again in nearly the same order, and then let us hope to be able to do these justice.

# JUNE DAYS.

'Twas one of the charmed days,
When the genius of God doth flow,
The wind may alter twenty ways,
A tempest cannot blow;
It may blow north, it still is warm;
Or south, it still is clear;
Or east, it smells like a clover-farm;
Or west, no thunder fear.
— EMERSON — *Wood Notes.*

The charm of the June days is due, in no small part, to the obliquity of the ecliptic. The inclination of the earth's axis at an angle of nearly twenty-three and a half degrees from the perpendicular to the plane of its orbit is a most important fact in mathematical geography, and yields as its result, when combined with the rotation of the earth on its axis and its revolution around the sun, that wonderful variety in the aspects of Nature which is the characteristic of life on the earth, especially in the Temperate Zones. The zones of climate are determined by it; the tropics and the polar circles owe their existence and present location to it. A difference of a few degrees, more or less, in this angle would have made the history of the world different. Civilization appears to depend upon it. Organic life, vegetable and animal, is deeply affected by the apparent northward or southward progress of the sun. Buds swell and open at its coming, and the birds, the welcome, joyous birds, herald its return. Scarcely has the sun crossed the Equator on its northward journey before the migratory birds are with us, and by the time it has reached the Tropic of Cancer they are all in their summer homes and are brooding over their nests, impelled by that mysterious inherited memory or instinct which leads them to do as their ancestors have done for ages, no one knows how long. When the sun has reached the Equator on its southern journey the fall migrations are beginning, and long before it has reached the Tropic of Capricorn the last of the summer visitants has bidden us adieu. They came arrayed in their gayest colors and singing their sweetest songs: they have gone with faded colors and almost in silence. We know they will come back, but we do not await their return with the same reverent feeling with which our ancestors in the old time awaited the coming of spring. The flight of the seasons affects us but little. We are as comfortable in winter as in summer, thanks to the improvements with which an advancing civilization has enriched our homes.

We no longer personify light and darkness, or think of the contest between summer and winter as worthy of our attention. But to our ancestors, more directly in contact with the daily and yearly phenomena of Nature, and more dependent on them for food and warmth and life, these were the great facts of existence, and this struggle of the powers of light with those of darkness was the foundation of their mythology and their creeds, and powerfully affected their lives.

The tide of Life which has been advancing all through the spring reaches its highest point in June. April and May are months of anticipation; their faces are turned toward the future glories. June is a month of realization; summer is here. The little modest spring flowers have departed; the half-opened leaves have expanded; the many-colored hues of the spring foliage have ripened into a nearly uniform darker verdure.

This is the most tuneful month of the year. The summer songsters have joined their music to the warbling of the earlier comers. Every morning opens with a pæan of praise; every evening closes with an anthem of peace. One of the most common of these songsters is the robin (*Merula migratoria*, Sw. and Rich.). Wilson Flagg in his charming "The Birds and Seasons of New England" pays this deserved tribute to the robin's musical ability: "I shall not ask pardon of those critics who are always canting about musical 'power' . . . for assigning the robin a very high rank as a singing bird. Let them say in the cant of modern criticism, that his performances cannot be great because they are faultless. It is enough for me that his mellow notes, heard at the earliest flush of dawn, in the busy hour of noon, or in the stillness of evening, come to the ear in a stream of unqualified melody. . . . The robin is surpassed by some other birds in certain qualities of song. The mocking-bird has more 'power,' the red thrush more variety, the bobolink more animation; but there is no bird that has fewer faults than the robin, or that would be more esteemed as a constant companion, —a vocalist for all hours, whose strains never tire and never offend. . . . The notes of the robin are all melodious, all delightful, loud without vociferation, mellow without monotony, fervent without ecstasy, and combining more of sweetness of tone, plaintiveness, cheerfulness, and propriety of utterance than the notes of any other bird. The robin is the Philomel of morning twilight in New England and in all the northeastern states of this continent. If his sweet notes were wanting, the mornings would be like a landscape without the rose, or a summer-evening sky without tints. He is the chief performer in the delightful anthem that welcomes the rising day. Of others the best are but accompaniments of more or less importance. Remove the robin from this woodland orchestra, and it would be left without a *soprano.*" This is high praise, but no one who is familiar with the facts will say that it is not deserved. And when we consider the wide distribution of this bird over all parts of North America, from Greenland and the islands of Bering's Sea to Cuba and Guatemala, we can imagine the vastness of that chorus which ascends from the voices of these birds in the interval between the earliest dawn and sunrise from April to July.

Our love for the robin, dating back to early childhood, does not prevent a kindred feeling for some other birds with which we have most pleasant associations. Something would be missing out of life if we could not yearly renew acquaintance with the little chipping sparrow or hair-bird (*Spizella socialis*, Bonaparte). He used to be the most familiar bird of thirty years ago, until the arrival of the English sparrow. If he is not attractive by his color, ashen-brown above and grayish-white beneath, wearing a velvety-brown skull-cap which easily distinguishes him from his kindred, his diminutive size and his marked sociability make him a favorite. He has, too, his own characteristic song, a long-drawn trill, heard early and late, a real and valuable contribution to the bird music of spring. If in my callow days I threw stones at these confiding sparrows I was never unfortunate enough to hit one, and I hope the recording angel blotted out with a tear the record of my misdeeds, as he did that of "my uncle Toby."

The catbird (*Mimus carolinensis*, Gray) was another of the familiar birds of those early days. It is hard to explain the prejudice against him. All the boys imbibed so readily the idea that it was a perfectly proper, even a commendable, act to stone every catbird and to rob all their nests that it seems now as if the idea must have been inherited. Dr. Coues says of him in his "Birds of the Colorado Valley": "For myself, I think the boys are right. Like many of the lower animals, they are quick to detect certain qualities, and apt to like or dislike unwittingly, yet with good reason. The matter with the catbird is that he is thoroughly commonplace. . . . The catbird has certainly a good deal to contend with. His name has a flippant sound, without agreeable suggestiveness. His voice is vehement without strength, unpleasant in its explosive quality. His dress is positively ridiculous—who could hope to rise in life wearing a pepper-and-salt jacket, a black velvet skull-cap, and a large red patch on the seat of his pantaloons?" Who does not recognize the truth of this picture? "Let it be the humble country-home of toil, or the luxurious mansion where wealth is lavished on the garden—in either case the catbird claims the rights of squatter sovereignty. He flirts saucily across the well-worn path that leads to the well, and sips the water that collects in the shallow depression upon the flagstone. In the pasture beyond, he waits for the boy who comes whistling after the cows, and follows him home by the blackberry road that lies along the zigzag fence, challenging the carelessly thrown stone he has learned to dodge with ease. He joins the berrying parties fresh from school, soliciting a game of hide-and-seek, and laughs at the mishaps that never fail when children try the brier patch." But it is in the mating season that he is heard at his best. "The next we see of the bird, he is perched on the topmost spray of yonder pear tree, with quivering wings, brimful of song. He is inspired; for a time at least he is lifted above the commonplace; his kinship with the prince of song, with the mockingbird himself, is vindicated. He has discovered the source of the poetry of everyday life."

A bird almost as familiar as any of these is the red thrush, as some writers prefer to call him, or the brown thrasher (*Harporhynchus rufus*, Cab.), a name and a bird both endeared to me by long and intimate acquaintance. "There goes a brown thrasher!" used to be a cry that would set a troop of boys gazing after one of their pet birds. Unconsciously, as boys do learn so many things, they had learned to distinguish his fine song in the open pasture or on the edge of the woodland, and to respect the nest with its treasure of four or five eggs or young ones. As the old Greek poet invoked the return of the swallow, singing "Come, come, O swallow, bringing the pleasant hours," so I welcome the return of this favorite songster, without whose presence the bright days of June would seem to lack somewhat of attaining their perfection.

It would be pleasant to dwell longer on the charm of birds, of the purple finch (*Carpodacus purpureus*, Bonaparte), of the yellow-bird (*Chrysomitris tristis*, Bonaparte), of the blue-bird (*Sialia sialis*, Baird), of the indigo-bird (*Cyanospiza cyanea*, Baird), of the Baltimore oriole (*Icterus Baltimore*, Daudin), whose brief, loud notes will be heard from the tree-tops, of the redstart (*Setophaga rutacilla*, Sw.), of the Maryland yellow-throat (*Geothlypis trichas*, Cab.), and of the many others which are to be found in glade and mead. One of them I cannot forbear to mention, the bobolink (*Dolichonyx oryzivorus*, Sw.). Not a singing bird in New England enjoys the notoriety of the bobolink. He sings only in the open field, where he can be seen as well as heard. He is almost continually in motion, singing on the wing as if in ecstasy. I have often seen one drop to the ground in the middle of a meadow where the grass was yet uncut, but although searching carefully I could not find a nest, so skilfully was it concealed. Bryant's "Robert of Lincoln'" is a well-known poem which has won its way to many hearts from its deep insight into, and sympathy with, the bobolink's nature; Wilson Flagg, a devout lover of Nature and author of two of the most interesting books on outdoor life in New England, especially in Massachusetts, "The Woods and Byways of New England" and "The Birds and Seasons of New England," has expressed the character of the bobolink's song most finely. Of the four stanzas of the little poem I venture to quote two in the hope that the whole may become better known.

### "THE O'LINCON FAMILY.

A flock of merry singing-birds were sporting in the grove;
Some were warbling cheerily and some were making love.
There were Bobolincon, Wadolincon, Winterseeble, Conquedle,—
A livelier set were never led by tabor, pipe or fiddle :—
Crying, 'Phew, shew, Wadolincon; see, see, Bobolincon
Down among the tickletops, hiding in the buttercups;
I know the saucy chap; I see his shining cap
Bobbing in the clover there,—see, see, see!'

. .  .  .  .  .  .  .

O what a happy life they lead, over the hill and in the mead!
How they sing and how they play! See, they fly away, away!
Now they gambol o'er the clearing,— off again, and then appearing;
Poised aloft on quivering wing, now they soar and now they sing,

'We must all be merry and moving; we must all be happy
and loving;
For when the midsummer is come, and the grain has
ripened its ear,
The haymakers scatter our young, and we mourn for the
rest of the year;
Then, Bobolincon, Wadolincon, Winterseeble, haste, haste
away!'"

The June flowers are rivals of the June birds in
interest. The roses are then in bloom, and the early
wild rose (*Rosa lucida*, Ehrh.), the swamp rose (*Rosa
Carolina*, L.) and the sweetbrier (*Rosa rubiginosa*, L.)
with their bright colors contribute to the beauty of the
woodlands. An interesting plant, not now found in
bloom within the city limits but not rare in the northern
part of the county and farther north, is the twin-flower
(*Linnæa borealis*, Gronov.). It gives an additional
interest to a plant to think that the name of a scientific
man, famous, world-famous perhaps, in his generation,
is linked forever with such an obscure memorial. The
men pass away, and a great part of what they accomplished is forgotten, but the little plants reviving with
every springtime keep their names fresh in the memories of students of plant life. *Linnæa* is named in
honor of the great Swedish naturalist, Linnæus, the
founder of modern botany, and appears, by the "Journal of a Tour to Lapland," to have been selected by

himself to commemorate his own name. Former botanists had named it *Campanula serpyllifolia*, but he, from a study of its structure, found it to constitute a new genus. He kept this idea in his own mind until he felt entitled to botanical commemoration, and then his friend Gronovius undertook to make it known to the world under its new name, to the gratification of Linnæus, who regarded it as a pledge of immortality. When Thoreau wandered through the Maine woods

"He saw beneath dim aisles, in odorous beds,
The slight Linnæa hang its twin-born heads,
And blessed the monument of the man of flowers,
Which breathes his sweet fame through the northern bowers."

I first found it in bloom one August day while walking along the highway from Bethlehem, New Hampshire, to Littleton. The air in the valley was still and filled with the sweet odor of pines, but at one spot there was a far sweeter odor which revealed the presence of the long-wished-for and late-delayed Linnæa. That road is remembered not so much for the two villages which it connects, as for being the one by the side of which grows one of the favorite northern flowers. It is pleasant to think, too, that this little plant grows over a wide area, in northern Britain, Lapland to northern Italy, cold and arctic Asia, as well as across America from New England to the far West.

Linnæus named in honor of a favorite pupil, Peter Kalm, a small genus of beautiful, exclusively American, plants. Kalm was destined by his family for the church, but after hearing a course of lectures by Linnæus, he determined to devote his life to the study of natural history. In the interest of the Royal Swedish Academy he traveled quite extensively in this country, especially in Pennsylvania, New Jersey, New York and Canada, about the middle of the last century, and sent home many new plants. Of the six known species of Kalmia, three are found quite abundantly in the county, even within the city limits, the pale laurel (*K. glauca*, Ait.), which comes into bloom early, the sheep laurel (*K. angustifolia*, L.), and the mountain laurel (*K. latifolia*, L.), perhaps the most conspicuous flowering shrub in June. Kalmia blossoms have a special adaptation for cross-fertilization that is worthy of notice. In the flower-bud each of the ten anthers is lodged in a small cavity, and the filaments are nearly straight. When the saucer-shaped corolla is opened, these filaments are curved outward and backward, resembling a curved spring. The anthers are not dislodged from the little cavities above mentioned unless disturbed. Humble-bees in search of nectar touch the filaments, which spring back with considerable energy and project the pollen from the two orifices at the apex of the anther-cells. Some pollen may be thrown in this way upon the single small stigma at the tip of the style, but more of it is thrown upon the under side of the insect's body, from which it is brushed off by contact with the stigmas of other flowers. Our woods present no more charming sight than a path lined on both sides with this handsome shrub in full flower. I have now in mind such a one. It should be seen on one of these rare June days. Far off through the waving branches overhead may be seen the pale blue sky flecked here and there with a thin streak of cloud, butterflies are flitting slowly by, the caw of a distant crow or the scream of a blue jay near at hand or the twitter of the smaller birds tells of the bird-life around; and on each hand are these beautiful white or pink flowers in such profusion that we scarcely know which to pluck or which to leave behind.

Kalm dedicated to Dr. Gaulthier of Quebec a genus of plants which embalms his memory, commonly called wintergreen or checkerberry or partridge-berry or boxberry (*Gaultheria procumbens*, L.). This is a peculiarly American plant, and is well known by its pleasant aromatic flavor, its shining evergreen leaves, its delicate flowers, and its scarlet berries. It is very generally distributed throughout New England, where it can enjoy the protection of evergreen as well as deciduous woods, and is about as well known to young persons as any of our native plants, not excepting the trailing arbutus.

Flowering shrubs hold a prominent place in the records of June. Thickly scattered along the roadsides and forming no inconsiderable part of the beautiful fringe with which Nature trims the edges of her robes are at least six species of the genus Cornus, the flowering dogwood (*Cornus florida*, L.), the dwarf cornel (*C. Canadensis*, L.), the alternate-leaved cornel (*C. alternifolia*, L. f.), the panicled cornel (*C. paniculata*, L'Her.), the round-leaved cornel (*C. circinata*, L'Her.), and the silky cornel (*C. sericea*, L.). The first of these is the most attractive. The flowers are borne in a head or close cluster surrounded by a large and showy four-leaved, corolla-like, white, sometimes pinkish, involucre. Scattered among the greenery it is admired by all who see it. Its bright red berries are no less ornamental. The dwarf cornel, often called bunchberry, rarely grows more than six or seven inches high. Its flowers seem like small copies of the larger species just described, and the connection is quite plain. In damp, cold woods to the northward it is very abundant.

The six species follow each other closely, so that for more than a month one or another of them is always in bloom. Each has its own characteristic habit of flowering, and is easily distinguished from the others.

They are members of an ancient family. In that strange scene described by Homer in the tenth book of the Odyssey, in which Circe changes some of the companions of Ulysses into swine, we read that, after the transformation: "Thus were they penned there weeping, and Circe flung them acorns and mast and fruit of the cornel tree to eat, whereon wallowing swine do always fatten."

The viburnums resemble the cornels somewhat when in flower, and some of them in fruit. Besides the hobble-bush which blooms early, we have in June these five species, the cranberry-tree (*V. Opulus*, L.), the sweet viburnum (*V. Lentago*, L.), the withe-rod (*V. cassinoides*, L.), the arrow-wood (*V. dentatum*, L.), and the maple-leaved viburnum (*V. acerifolium*, L.). The viburnums are easily distinguished from each other by their leaves, and two of them take their specific names from this character. They, too, like the cornels, succeed each other, so that for several weeks we can always find one or more of them in bloom.

June is a favorite month for many of the orchids. The arethusa (*Arethusa bulbosa*, L.), one of the most beautiful of the American wild flowers, the tway-blade (*Liparis liliifolia*, Richard), two of the pogonias (*P. verticillata*, Nutt.) and (*P. ophioglossoides*, Nutt.), at least two of the habenarias, *Habenaria Hookeri*, Torr.

and the purple-fringed orchis (*H. fimbriata*, R. Br.), and three of the lady's slippers, the stemless (*Cypripedium acaule*, Ait.), the large yellow (*C. pubescens*, Willd.) and the small yellow (*C. parviflorum*, Salisb.). These form a very interesting group, both in themselves and in their relation to a great family. They are much sought and are highly prized. They hold the same rank among our wild flowers that the scarlet tanager, the rose-breasted grosbeak, the bobolink, the Baltimore oriole, the redstart and the goldfinch hold among the birds.

The development of ferns is an interesting sight connected with our native flora. Each genus has its own manner. This shepherd's crook will develop by and by into a tall osmunda, of which we have three species. The name of this genus has given rise to some speculation as to its derivation. Some suppose it to be from Osmunder, the Celtic name of the god Thor of the Scandinavian mythology. Moore in his "Popular History of British Ferns" gives another, to which I refer the reader curious in such matters. In Hooker and Baker's "Synopsis Filicum" six species of this genus are described, three of which, the cinnamon fern (*Osmunda cinnamomea*, L.), the interrupted fern (*O. Claytoniana*, L.) and the flowering fern (*O. regalis*, L.), are cosmopolitan. The first is found from Newfoundland and

Canada to Mexico, in the West Indies, Guatemala, the United States of Colombia, and on the Organ Mountains of Brazil, also in Japan, Mantchooria and the Amur country; the second is found in Canada, Newfoundland, throughout the United States, and on the Himalayas to the height of ten thousand feet; the third is accredited to Sweden, Russia, Siberia, Japan, the Azores Islands, the Barbary States, the Himalayas, Bombay, Cochin China, Hong Kong, Zambesi-land, Natal, Cape Colony, and from Canada to Rio Janeiro; and yet the three species may often all be found here in a space less than ten feet square. The other three species of osmunda are all found on the eastern side of Asia.

We have representatives of at least a dozen other genera, all of which are crowding forward in June. One of our plants, akin to the ferns, and formerly counted with them, is the moonwort (*Botrychium Virginianum*, Swartz). This has a very remarkable geographical range. In Europe it is found only in Norway (*fide* Macmillan); but it abounds in the United States, on the mountains in Mexico, on the Raklang Pass of the Himalayas, and is abundant on the mountains of Australia and New Zealand, where it is boiled and eaten by the natives. The distribution of this plant over such widely-separated areas is a very puzzling problem.

There are some humbler plants which are no less interesting though they may be considered as weeds.

One of the smallest of these grows on the edge of gravel walks and opens its pink-red corolla during the middle of the day when the sun is shining. In the morning and in the late afternoon we might pass it unnoticed, but at midday when its many blossoms are open, small though they are, we should be attracted to it. I learned to know it under the name *Spergularia rubra*, Presl. but it is known also as *Buda rubra*, Dumort., and *Tissa rubra*, Britton. Whichever name may be finally accepted, the little plant will be not a jot the less interesting, nor bloom a whit the less profusely.

The ox-eye daisy (*Chrysanthemum Leucanthemum*, L.) and the cone-flower (*Rudbeckia hirta*, L.) are two of the June flowers conspicuous by their great numbers. Though they may be regarded as pestiferous weeds when encroaching upon cultivated lands, there is plenty of room for them in the waste lands everywhere; and the golden-yellow disks with the crown of silver-white rays of the former, and the dark brown disks with the crown of golden-yellow rays of the latter will not fail of admirers. Both of them are immigrants. The ox-eye daisy has followed the course of empire and comes to us from the old world; the cone-flower has reversed this course and comes to us from the prairies of the Mississippi Valley. Both of them have come to stay. They thrive in this land, and, if they could speak, would, no doubt, say "Our lot is cast in pleasant places: here will we abide."

No one who has seen a meadow full of the scarlet painted-cup (*Castilleia coccinea*, Spreng.) can forget the charm which such a scene casts over his sense of the beautiful. I know only one such meadow. It is a Mecca toward which I turn my face annually, and not alone. My friends may share the pleasure with me without diminishing my own. These scarlet tufts, glowing in the green of the grass, are almost like flashes of fire, and suggest the scarlet tanager and the Baltimore oriole as fit occupants of the orchard and woodlands near. The name of Castillejo, a Spanish botanist, cannot be forgotten as long as it is associated with this favorite flower. Like Linnæus, he has gained a sort of immortality.

Close by the painted-cup I find the pink azalea (*Rhododendron nudiflorum*, Torr.), a member of a great family whose homes are chiefly in temperate North America and among the mountains of India. When seen in the wildwood, with the surroundings in which Nature has placed them, our native species seem in no way inferior to the most highly prized exotics; when transplanted to the garden they sometimes lose that attractiveness by comparisons for which they were evidently not intended. Fortunate are we if we can find time and opportunity in these charmèd June days to catch a glimpse of these many beautiful creations of unassisted Nature. Great is Art; but there is One greater and older, by obedience to whose laws Art itself is able to attain its highest ideals.

*Rhododendron nudiflorum, Torrey.*
*The Swamp Pink.*

"*What splendid masses of pink, with a few glaucous green leaves sprinkled here and there—just enough for contrast.*"
—H. D. THOREAU.

# THE JULY PAGEANT.

> Morn in the white wake of the morning star
> Came furrowing all the orient into gold.
>
> . . . .
>
> It was the deep midnoon:—
> And at their feet the crocus brake like fire,
> Violet, amaracus and asphodel,
> Lotos and lilies.
>
> . . . . . .
>
> The charmed sunset lingered low adown
> In the red West.
> —Tennyson.

The clouds, like stately argosies, move slowly across the summer sky, and the cloud-shadows darken the green hillsides and the deep meadows, where the tall grasses are bending in long waves under the gentle breeze. Swallows, on unwearied wing, are floating in wide circles in the upper regions of the air or are skimming the flashing surface of the pool in the intricate mazes of their flight. A flock of redwing-blackbirds, "cong-ka-rees," in the thickets of button-bush (*Cephalanthus occidentalis*, L.), where the river in its sluggish course embraces the little islets. The chirping of myriad insects adds to the ceaseless hum which attends the onward march of summer.

The gates of life are now opened wide and the *Via Æstiva* is crowded with the gay throng that fills its every nook and corner and yet moves steadily forward to the accompaniment of Nature's music, the song of birds, the rippling of innumerable brooks, the murmuring of the wind among the branches, and the fluttering of multitudinous leaves.

There have been many famous processions in the world's history, the story of which, where preserved, the world loves to linger over. In the mind's eye we can still see Solomon, accompanied by the elders of Israel and all the heads of the tribes, the chief of the fathers of the children of Israel, going down to the city of David, which is Zion, to bring the ark of the covenant of the Lord, with the sound of trumpets and cymbals, unto Jerusalem, to the great and magnificent temple which he had there builded. Lilies still bloom in the fields of Palestine and they, too, have their glory. We seem to hear the Great Teacher, as He sits on the mountain side, saying to the multitude: "Consider the lilies of the field, how they grow; they toil not, neither do they spin; and yet Solomon in all his glory was not arrayed like one of these." We turn the pages of Pindar, and again the victors in the Olympic games are returning home in triumph, bearing the wreath of wild olive, having won for themselves and their native city an imperishable name. In imagination we may stand in the Forum at Rome and see some great proconsul returning victorious from a foreign war:—it may be Scipio Africanus fresh from the conquest of Carthage after the second Punic war, or Æmilius Paullus with Perseus, king of Macedon, in his train, or Caius Julius Cæsar, "the laurelled scholar, the sun-bright intellect," "the foremost man of all this world," celebrating four triumphs at once because his life has been too busy to celebrate separately his successes in Gaul, in Egypt, in Pontus and in Numidia:—they pass in splendor along the Sacred Way and up the Sacred Hill, offering in the magnificent display of silver and gold and precious stones and other treasures of the vanquished lands such gorgeous spectacles as the world has rarely witnessed. There have been coronation pageants and royal progresses, not a few, through the streets of capital cities; processions of the elect of a nation, like that of the States General of France at the beginning of the Revolution, so vividly described in Carlyle's pages, or like that, to which no pen but Macaulay's could do justice, at the opening of the High Court of Parliament which was to sit in judgment upon Warren Hastings. Yet, like Palmyra in the desert, nothing now remains of them but a name; like the baseless fabric of a dream they have all vanished into thin air; they cannot return.

But in the procession of Life, which passes along the Summer Way in this month of July — Did not the Romans change the name of their fifth month, Quintilis, to Julius, in honor of Julius Cæsar, and does not the very name, therefore, suggest a triumphal march? — we shall find the old familiar faces in the old familiar places. It is ever old, and yet it is ever new. It does not lose its charm and attractiveness for the thoughtful observer. It is itself constantly changing, but there is an element of certainty about it on which we can put a firm trust. In a year we hope to hear again "The Voice of the Grass" as it whispers:

> "Here I come creeping, creeping everywhere,
>     By the dusty roadside,
>     On the sunny hillside,
>     Close by the noisy brook,
>     In every shady nook,
> I come creeping, creeping everywhere."

When the heat of July has succeeded to the balmy days of June, and the green pastures are turning brown through the ripening of the grasses, the vernal flowers have mostly disappeared to give place to those of the midsummer and the autumn. The season of hay-making has come. The rattle of the mowing-machine in the open meadows and the whetting of the scythe by the busy mowers on the rocky upland or along the walls and fences lined with thick shrubbery blend pleasantly with the other sounds which now fall upon the ear. The tall butter-cups, the purple cone-flowers, the ox-eye daisies, the meadow rue and many others fall with their more modest and humble neighbors, the grasses, with no Burns to sing their fate. The fragrance of the new-mown hay, the last tribute of the dying flowers, is associated with many of the most pleasant rural scenes. The grasses form a very conspicuous part of the summer vegetation by their size, their number, and the variety of the species. The smallness of their flowers may deter the young botanist from attempting to study them, but no family repays study better. Rare grasses are as satisfactory "finds" as any other plant can be. When Robert Dick, a baker at Thurso in northern Scotland, and at the same time an enthusiastic geologist and botanist, discovered the northern holy grass (*Hierochloë borealis*, R. & S.) growing near Thurso, the botanists of Great Britain were greatly surprised, and the Royal Botanical Society sent Dick a special vote of thanks for his paper announcing the locality of the grass and for specimens of it. Years before, this plant had been admitted into the British Flora on the authority of Don;

but no one else had found it, so that after a time it was placed on the doubtful list and then finally dropped out altogether.

The Grass Family is one of the royal families of plants, and one of the largest, probably being surpassed in the number of species by the Compositæ and Leguminosæ only. It is also widely diffused over all the habitable parts of the globe, having no limits except those which bound vegetation in general, reaching the outposts of flowering plants both in the polar regions and near the snow-line on high mountains. About three thousand five hundred species have been already discovered and described, some of which are characteristic of the tropics, some of Alpine heights; but no part of the world is more highly favored in the variety and social character of its grasses than the north temperate zone, where, owing to a uniform rainfall, the grasses unite to form extended areas of turf covering large meadows. In tropical regions the turf disappears, and the grasses become larger, more isolated like other plants, fewer in the number of individuals, but with broader leaves and more showy flowers.

The aspect of Nature is determined largely by this family. The great grass regions form characteristic parts of each of the grand divisions, such as the great steppes of central and western Asia, the great eastern plain of Europe, the plains of central Africa, the llanos and pampas of South America and the prairies of North America.

About seven hundred species of grasses are found in the United States, and about seventy in Worcester County, some of which, of course, are cultivated for food of cattle. In the early history of this country, particularly in the northern states, while the settlements were sparse, the natural pasturage was abundant, and the natural meadows and marshes furnished a supply for winter feeding. But in the course of time, by the increase of population, the farms began to crowd each other, and the range for cattle was restricted. Then began the cultivation of the four principal grasses, herd's grass (*Phleum pratense*, L.), orchard grass (*Dactylis glomerata*, L.), Kentucky blue grass (*Poa pratensis*, L.), and redtop (*Agrostis alba*, L.). We can realize the extent of the culture of these and other grasses when we are told that the haycrop of the United States in a single year amounts to more than forty-six million tons from thirty-eight million acres, and is worth in round numbers about four hundred million dollars; and this, in addition to the amount consumed by grazing animals.

Besides the commoner forms of grasses, which we all readily recognize as such, there stands preëminent a group of cultivated grasses, through which this family contributes more to the sustenance of man and beast than all other plants together. They are the staple cereals of the world, cultivated from time immemorial, wheat, rye, oats, barley, rice, millet, maize. There can be no doubt that they were originally selected from wild forms on account of the size, quantity and nutritive value of their seeds. And when this fact of their value was discovered, the discovery would soon follow that, by planting these seeds in suitable ground and caring for the growing plants to the exclusion of all other vegetation, a certain and reliable source of food would be obtained; and here would be a beginning of agriculture, with capacity for infinite development.

Wheat, rye, oats, barley, millet and rice are forms belonging to the old world, but now widely scattered over the new. The new world has, in return, given some gifts to the old, prominent among which are maize, the potato and tobacco, but the greatest of these is maize. With an acreage in the United States alone of seventy-eight million acres it is the largest arable crop grown in any country. Its overshadowing influence in our agriculture is shown by the fact that the area devoted to its culture in many districts exceeds that given to the special crop for which that district is famous. The eleven states of our cotton belt devote, as a whole, a larger area of their cultivated land to corn than to cotton. It is so in the great wheat belt. In measured quantity our crop of a single year has excceded the wheat crop of the whole world. Plainly, corn is king among the princes of this royal family. What more fitting emblem for the great Republic than this kingly plant whose grain is of the color of gold, which is grown to some extent in every state and territory of our Union, and in almost every county in which agriculture is carried on!

In southeastern Asia, the hive of the human family, three other grasses of great economic value probably had their origin, rice, sugar-cane and the bamboo. Rice is used for food by more people than any other one grain. The mere enumeration of the uses of the bamboo would fill many pages.

What a variety of products comes from these few members of the grass family — flour, meal, corn, oats, barley, rye, sugar, molasses. Fermented and distilled liquors are no insignificant part of the product. Sugar, molasses and rum are three very unlike products, yet they are all obtained from the stalk of the sugar-cane. What countless springs of activity are set in motion by this one family! Agriculture, manufacturing and commerce are fostered by it.

Our humbler grasses will certainly take on an added interest when we consider their kinship with these noble forms. Certainly, in the upland pasture we shall still look for the quaking grass (*Briza media*, L.), favorite among our local grasses; in the meadows the oat-grass (*Arrhenatherum avenaceum*, Beauv.) will still please with its slender, graceful panicle; close by, the velvet-grass (*Holcus lanatus*, L.) will attract by its pale color and soft-downy appearance; on the edge of the thicket where we first found it a dozen years ago or more, we shall still hope to find *Brachyelytrum aristatum*, Beauv.

In their structure the grasses form an isolated family, showing close relationship to the sedges only, but differing from them in the structure of the fruit and the embryo. About eighty species of sedges have been thus far discovered in the county. Unlike the grasses, the sedges are of little use for food or in the arts. Mingled with grasses, for which they are easily mistaken, the sedges are sometimes eaten by cattle, but are lacking in those valuable properties which render the various grasses so useful to man. Lacking those qualities which attract attention to the more favored family, these perform, we are warranted in believing it, no mean part in the economy of Nature. In some parts of the world, especially in Europe, they are of great use in binding the sands of the sea shore so that the strong sea gales cannot blow the sand of the dunes inland.

Some of the sedges are among the earliest flowers of spring, others fill up the long summer days, few linger into the autumn. They are mostly distinguished from each other by the characters of the fruit rather than of the flowers. They form an interesting group for study, and more than one monograph has been devoted to them. The most famous member of the group taken as a whole is the papyrus plant, which is often a conspicuous feature of African vegetation. One species was commonly used in Egypt for the purposes of writing, and was, in fact, the paper of the period. Its use as paper continued until the twelfth century, when it was superseded by parchment and by paper made from rags. By the discovery of papyrus rolls in Egypt in recent years some valuable works of the classical Greek authors have been rescued from oblivion.

Of the genus *Cyperus*, which is largely represented in the tropics, we have four species. Their common book-name is galingale, a name borrowed from over the sea. In *The Lotos-Eaters* Tennyson has chosen the galingale along with the palm as symbols of that dreamy land to which Ulysses came in his wanderings and from which he so soon hurried away, lest he and his men should lose their love of home and country, of wife and child.

> "The charmed sunset linger'd low adown
> In the red West; thro' mountain clefts the dale
> Was seen far inland, and the yellow down
> Border'd with palm, and many a winding vale

And meadow, set with slender galingale ;
A land where all things always seem'd the same !"

By the margin of the pond where we find *Cyperus dentatus*, Torr. we may, perhaps, see one of those transformations which form an interesting subject of study in the animal world. This is a suitable spot to watch the dragon-flies, great and small, as they dart over the water, flashing with their beautiful colors. The dragon-fly, as well as the butterfly, has three stages of growth, after being hatched from the egg. At first it lives in the water as a larva or grub, developed from one of a series of eggs laid by a dragon-fly on a leaf of some water plant. It is chiefly remarkable for its masked mouth and for the power of moving by means of a jet of water expelled from the tail. It spends most of its life in this state, crawling about on the bottom of the pond and feeding upon other aquatic insects. After some months it attains its full size, having changed its skin many times, and is now in the pupa state, which differs from the larval state chiefly in being larger and in developing rudimentary wings. In this state it is just as active and voracious as the larva was, and leads a similar life. But by-and-by the time arrives for the final transformation. The pupa crawls slowly up the stem of some water plant which reaches above the surface. Gradually the outer skin splits and the perfect insect disengages itself and works its way out, having now wings and other organs like its parent of nearly a year before. In a few hours the wings become expanded and hardened, and the brilliant colors gradually become apparent. In *The Two Voices* Tennyson gives us a picture of this scene.

> "To-day I saw the dragon-fly
> Come from the wells where he did lie.
> An inner impulse rent the veil
> Of his old husk: from head to tail
> Came out clear plates of sapphire mail.
> He dried his wings; like gauze they grew;
> Thro' crofts and pastures wet with dew
> A flash of living light he flew."

During the single month of their matured life dragon-flies haunt the ponds and streams, preying on butterflies and moths and other insects. Sometimes we see them hovering over, or settling upon, the mud left by summer showers. They are classed with the insects beneficial to agriculture, because their chief mission is to destroy insects which feed on vegetable life.

"There is but one butterfly," says Scudder (Am. Nat., vol. X., 392), "whose range is so extended as to merit the name of cosmopolitan; it is the Painted Lady or *Vanessa Cardui*, L. (*Cynthia Cardui*, Fab.). With the exception of the arctic regions and South America it is distributed over the entire extent of every continent." We are not surprised, therefore, to find it hovering over these wayside thistles (*Cnicus lanceolatus*, Hoffm.). It is one of our favorite butterflies, being associated in our thoughts with its kindred *Cynthia Huntera*, Fab., *Vanessa Antiopa*, L., *Nymphalis Arthemis*, Drury, and the rare White Mountain butterfly (*Hipparchia semidea*, Say), which we once saw on the summit of Mount Washington a few years ago. The butterflies, both by their numbers and by their beauty, add something to the charms of summer which we cannot fail to appreciate.

Although most of the trees and shrubs bloom early, there are yet a few which seem to linger so that each month may have its share of them. Largest of these is the chestnut (*Castanea sativa*, Mill., var. *Americana*, Gray) conspicuous far and near by its long yellowish catkins, relieved by the dark green foliage, attractive in flower and in fruit. At first sight the woods, when these trees are in flower, seem to be composed almost entirely of chestnut-trees, but a closer inspection shows there is a wonderful variety. In the open pasture where it has had a chance to grow at will, the chestnut is often a stately tree.

All our sumachs may be found in bloom this month, although some of them first open their flowers in June. At their head comes the poison ivy (*Rhus*

*Toxicodendron*, L.) with its leaf composed of three leaflets, a fact which easily distinguishes it from the woodbine or Virginia creeper (*Ampelopsis quinquefolia*, Michx.), which has five leaflets for its normal number, although six and even seven are often found. Then comes the stag-horn or velvet sumach (*Rhus typhina*, L.), easily distinguished by its densely velvety-hairy branches, the tallest of the genus in New England. The poison dogwood (*Rhus venenata*, DC.) is found in bloom about the same time. It is an elegant shrub, with a characteristic stem and foliage, the dread of those liable to be affected by its touch. Following closely is the smooth sumach (*Rhus glabra*, L.) easily distinguished by its smooth leaf-stem; and about a fortnight later the dwarf sumach (*Rhus copallina*, L.), of smaller size, of a smaller number of leaflets, and with a winged petiole, brings up the rear.

Along country roads where they pass through low lands or by the margins of swamps or ponds, the characteristic inflorescence of the button-bush, the perfectly globular shape of its heads of nearly white flowers, attracts our attention. Almost always growing in the water, it defies attempts at cultivation. The swamp birds love it and often build their nests in it. The sweet pepperbush (*Clethra alnifolia*, L.) delights in similar situations. It is by no means to be despised when we see it putting forth its long racemes of fragrant white flowers from the midst of the thick dark foliage which lines the margin of the wooded swamp.

What plant is this from the sunny south with petals dyed in bright purple? It is our one northern species of the one northern genus of the large tropical order Melastomaceæ. It is the meadow beauty (*Rhexia Virginica*, L.), fit associate for the yellow lily (*Lilium Canadense*, L.), and the wood lily (*Lilium Philadelphicum*, L.), and the yellow fringed orchis (*Habenaria ciliaris*, R. Br.), or whatever else is brightest and best among the July flowers.

Another northern representative of a tropical order is the pipewort (*Eriocaulon septangulare*, With.), found quite abundant in the eastern United States and on the Isle of Skye and on the west coast of Ireland, and nowhere else in Europe. It has an interesting connection with the history of botany. It fixed the destiny of Robert Brown, who, upon the completion of his medical studies, was attached as ensign and assistant to a Scotch militia regiment stationed in Connemara. This inconspicuous plant, with which he then became acquainted, caused his life to be directed to the exclusive service of botany. Accompanying a recruiting party of his regiment to London in the summer of 1798, and visiting his friend, Dr. Withering, near Birmingham, he was induced to introduce himself to Dr. Dryander, librarian to Sir Joseph Banks, and to show him his researches upon this plant. The favor of Sir Joseph Banks was thus obtained for him, and he was welcomed as a regular guest at his house during a five months' stay in London, and was soon proposed to the government as naturalist of the exploring expedition to New Holland under Captain Flinders. His career was thus determined; and on his return from Australia the wonderful sagacity and insight which he showed in his investigations of the rich and peculiar vegetation of that new region gave him the eminent position which he retained for more than half a century.

I have found it growing to the height of three or four inches on the sandy shore at the upper end of Lake Quinsigamond, and in water two feet deep at the southern end of the Lake, where the flower-heads, on stems of needed length, reach just above the surface, a case of adaptation to circumstances that seems to imply something more than mere vegetative life. Not far away from this sandy shore and in other moist places two of the sundews (*Drosera rotundifolia*, L. and *D. intermedia*, Hayne, var. *Americana*, DC.) may be found. The contrivance by which they catch insects and their method of disposing of those that are caught have been the subject of much study.

There are others of the plants of July which, as they pass along in an unbroken succession, cannot fail to claim for themselves more than a passing notice.

But the humblest of them all is as great a mystery as the proudest. Within each one of them Force is working; there is life in each one of them. They are responsive for a time to heat and light and moisture, until they have provided for the continuation of their species, and then they die. As long as there is life there is motion, change. From the old forms new forms are rolling out or evolving; and our conviction is deepening that one purpose runs through the ages, and that the world is emerging into a broader day.

The study of these humble plants may lead us to an apprehension of Order in Nature; it may at least change our faculty of sight into an art of seeing.

> "Flower in the crannied wall,
> I pluck you out of the crannies;—
> Hold you here, root and all in my hand,
> Little flower — but if I could understand
> What you are, root and all, and all in all,
> I should know what God and man is."

# THE AUGUST FIELDS.

> Who can see the green earth any more
> As she was by the sources of Time?
> Who imagines her fields as they lay
> In the sunshine, unworn by the plough?
> Who thinks as they thought,
> The tribes who then roam'd on her breast,
> Her vigorous, primitive sons?
> —MATTHEW ARNOLD—*The Future.*

Golden-brown is the color of the August fields as they lie basking in the glowing sun. Whatever may have been the beauties or charms of the earlier months, August is the golden age of the year. The earliest spring days looked forward to it for the fulfillment of their promise. Bright, indeed, were the days that are no more, but they were not the brightest. Their glories have grown dim and have faded away in the light of midsummer. We may long for the charms of those early days and may think of them as the golden age of the year, but the golden age of the year as of the world is not in the Past but in the Future.

In the introduction to his learned work on "The Development of the Feeling for Nature among the Greeks and the Romans" Dr. Biese says: "If we in our knowledge-proud time look around upon the result of modern thought and modern activity, and see with admiration in all departments of human endeavor new ideas rule and vast revolutions ever accomplished or prepared, then it will seem to us as if a whole world separates us from the Past of the earlier centuries, as if our entire mode of viewing things were a totally changed one which is only in the remotest degree like anything of earlier times. And on the other hand, in the midst of the fermentation, distraction and unrest of modern life and strife, the feeling of sadness seizes us, as if the Past had possessed something which we are now deprived of, as if we had lost the irrecoverable; and this longing weaves its magic mantle about a world of days long dead, in which the unsatisfied soul strives to find realized all which it misses so painfully in the Present. These two feelings, of pride over the immense progress of modern thought in comparison with the Past and of pain that a happy time born from the harmony of the outer and the inner life has vanished, hinder only too easily an objective appreciation of classic antiquity."

There have been no centuries in the world's history that have felt a more wide-spread and deeper interest in the appearances of Nature than the present and the last.

"In simple states of culture the feeling for Nature may appear hearty and subtile and have the charm of unconscious simplicity," continues Biese, "but the perfect development into true manhood which is built on the foundations of a higher culture makes man susceptible to the plastic power of Nature. In different stages of culture Nature will make different impressions upon men."

Long ago Sophocles, thinking of the gentle Antigone, stood by the blue Ægean and heard the grating roar of the pebbles which the waves, driven by blasts from the Thracian coasts, drew back and flung at their return up the high strand, and there came into his mind the thought of the turbid ebb and flow of human misery that creeps to generations far. Matthew Arnold, thoroughly imbued with the spirit of the old Greek culture modified by the powerful influences of the best modern training and thought, has looked from his window at Dover out over the straits to the cliffs near at hand and the gleaming light on the French coast, and to his mind has come a picture of the Sea of Faith once at the full but now

> "Retreating to the breath
> Of the night-wind, down the vast edges drear
> And naked shingles of the world."

While he can say his special thanks are due to Sophocles

"Whose even-balanced soul,
From first youth tested up to extreme old age,
Business could not make dull, nor passion wild;
Who saw life steadily, and saw it whole;

yet in the pure June night, in the heart of the English Lake-district, as he thinks of Wordsworth, who

"Was a priest to us all
Of the wonder and bloom of the world,
Which we saw with his eyes, and were glad,"

and as he questions mountain and shadow and lake, whether the charm, the grace, the beauty, the romance that we feel are in them or in the poet's voice, which reveals what they are, he can hear the voice of Nature herself reply:

"Loveliness, magic and grace,
  They are here! they are set in the world,
  They abide; and the finest of souls
  Hath not been thrill'd by them all,
  Nor the dullest been dead to them quite.
  The poet who sings them may die,
  But they are immortal and live,
  For they are the life of the world."

He, too, realizes that

"Nature is fresh as of old,
  Is lovely;"

and that, when our eyes appear blinded, it is only some dark shadow that has interposed. And the thought comes to him that, despite the gloom and the sorrow, the care and the turmoil that brood over life:

> "Haply, the river of Time—
> As it grows, as the towns on its marge
> Fling their wavering lights
> On a wider, statelier stream—
> May acquire, if not the calm
> Of its early mountainous shore,
> Yet a solemn peace of its own,"

and the thought brings renewed hope and comfort to his desponding heart.

The long lists of outdoor books, which deal with the picturesque or æsthetic side of Nature as distinguished from the scientific or the practical, would seem to indicate an increase in the number of those who are able to appreciate or wish to learn to appreciate the charms in this study of Nature. As the reading of Robinson Crusoe has incited many a youth with a longing for the sea and the strange adventures incident to a mariner's life, so the reading of a good outdoor book has been a guide and a stimulus to a truer love for outdoor life. Such books may not be considered as literature in the same sense as Homer and Plato, Dante and Boccaccio, Shakspere and Addison, Irving and

Bryant, nor is the reading of them considered a necessary part of a liberal education, yet some of them have lived long enough, and with a sufficient fame, to be entitled to the rank of classics.

As there were many brave men before Agamemnon, and many poets before Homer, so there were many anglers before gentle Izaak Walton, who loved to sit quietly, in a summer's evening, on a bank a-fishing, and who, in thought, if not in word, uttered the prayer of Jo. Davors, Esq.:

"Let me live harmlessly, and near the brink
   Of Trent or Avon have a dwelling-place,
Where I may see my quill or cork down sink
   With eager bite of perch, or bleak, or dace;
And on the world and my Creator think;
   Whilst some men strive ill-gotten goods t' embrace
And others spend their time in base excess
   Of wine, or worse, in war and wantonness.

Let them that list, these pastimes still pursue,
   And on such pleasing fancies feed their fill;
So I the fields and meadows green may view,
   And daily by fresh rivers walk at will,
Among the daisies and the violets blue,
   Red hyacinth and yellow daffodil,
Purple narcissus like the morning rays,
   Pale gander-grass and azure culver-keys."

## THE AUGUST FIELDS.

From "The Compleat Angler" of Izaak Walton in 1653 there is a long interval to "The Natural History of Selborne" by Gilbert White in 1789 and "Forest Scenery" by William Gilpin in 1791, but let us not suppose for a moment that the interval was a barren one. The "forgotten worthies" form a noble roll when their names are rescued from oblivion. The last fifty years, especially the last twenty, have been prolific in books of this class, and the demand has been apparently equal to the supply. Thoreau, Starr King, Higginson, Wilson Flagg, Burroughs, Torrey, Bolles are some of the American names, and Gosse, Kingsley, Darwin, Macmillan, Taylor, Hamerton, Heath, Worsley-Benison, Knight and Jefferies are a few of the English names which at once recur to the mind as preëminent among the writers of this class. More than ever before do the novelist, the traveler and the poet endeavor to paint with fitting words the phenomena of external Nature, as a setting for the thought they wish to convey.

The August fields are rich in flowers belonging to the great order Compositæ or compound flowers. A few of them, like the dandelion and the daisy, bloom early, but most of them belong to the late summer and autumn. They are eminently social flowers. They are distinguished from simple flowers by the following characters: there must be flowers collected into a compound head; five anthers—occasionally four—growing together around the single pistil, with filaments separate; and a single seed to each flower. In the lobelias the anthers will be found growing together, but the flowers are not in heads; in the teazles the flowers are in heads but the anthers do not grow together, so that a slight examination will suffice to distinguish any member of this great order, which contains at least ten thousand species. It includes plants of all habits, annual, biennial and perennial, herbaceous and arborescent. They are met with all over the globe, forming varying proportions of the flora in different regions. In the island of Sicily they form one-half of the known species native to the island, in France about one-seventh, in North America about one-sixth. They thrive in all soils; some where they are drenched daily by the sea, some on dry uplands, some in the forest, some by the roadside, some in meadows. The herbaceous forms vary in height from three or four inches in the small everlastings and dwarf dandelion to six or eight feet in the sunflower of the gardens. In Chili many of the species of this order are shrubs, on the lonely island of Saint Helena some of them are trees.

This order furnishes a storehouse of species for the florist. They swarm in every garden; scarcely a mixed boquet is complete without them. Asters, chrysanthemums, daisies, marigolds, dahlias, with hosts of others with no common names, belong here, all full of beauty and all favorites with some one.

Almost hiding the old stone walls of the pastures, lining the country roadsides, meeting us in the depths of the woodland, but delighting especially in the moisture of the swamps, first in number and almost first in size are the golden-rods, with their nodding yellow plumes. The technical name of the genus is *Solidago*, from the Latin words *solidus* and *ago*, *I make whole*, in allusion to reputed vulnerary qualities. In an afternoon ramble in this vicinity we ought to be able to find easily nine or ten species, among which might be several well-marked ones, the sweet golden-rod (*S. odora*, Ait.), the smooth (*S. cæsia*, L.), and the white (*S. bicolor*, L.).

> Born in the prime of the year, the pride of the heart
>     of the summer,
> Filling the field and the roadside with beauty too
>     often neglected,
> Cheering the heart of the man and the child with
>     the wealth of thy color,
> Golden-rod, child of the sun, adorned with thy father's
>     own glory,
> Dear art thou to the lover of Nature, who knows how
>     . to prize thee,
> Common though thou mayest be, but fresh from the
>     hand of the Maker.

Closely associated with the golden-rods are three species of *Eupatorium*. The largest of them is the purplish one, sometimes called by its book name, Joe-

Pye weed or trumpet-weed (*E. purpureum*, L.), which varies from two to twelve feet in height, and seems to fill, in low and moist situations, nearly all the space left vacant by the golden-rods, and even to strive with them for supremacy. Every gatherer of herbs is familiar with the second, well known as thoroughwort (*E. perfoliatum*, L.), and readily distinguished by the fact that the leaves are opposite to each other along the stem and grow together by their broad bases, encircling it, and by its clusters of small white flowers. The handsomest of the three is the white snake-root (*E. ageratoides*, L.), the flower of which very much resembles the *Ageratum* of the gardens. The three species are very conspicuous, both in size and in color, and play no insignificant part in forming the great masses of color which are the glory of the August flowers.

More than a dozen species of asters, white, blue, and purple, large and small, add to the beauty of the flora of this month and the next. They are so numerous that common names have not been given to them. Vying with the golden-rods in abundance, and surpassing them in the size of the heads and the variety of colors, the asters are among the most highly prized of the late summer and autumn flowers. My favorites among them are *Aster patens*, Ait., *Aster lævis*, L., *Aster Novæ-Angliæ*, L., *Aster salicifolius*, Ait., and *Aster linariifolius*, L.

By the dusty roadside as well as along the woodland path we come upon the trailing or erect stems of the bush-clover, of which five species are quite common, especially *Lespedeza polystachya*, Michx., and *L. capitata*, Michx. Closely associated with them are some of our nine species of tick-trefoil, *Desmodium*, of which *D. Canadense*, DC., and *D. nudiflorum*, DC. are the most noticeable.

Sproutlands are favorite places for the false foxglove, *Gerardia*, of which we often find three species near each other, *G. pedicularia*, L., *G. flava*, L., and *G. quercifolia*, Pursh, with large yellow flowers, almost covering the tall, more or less branching, and leafy stems. These are showy plants, and attempts have been made to cultivate them, but being partially parasitic on the roots of other plants, they have defied all such attempts. In the low lands by the roadside or on the gravelly margin of some pond, we may find the loveliest, though smallest, of them all, the purple gerardia (*G. purpurea*, L.).

Shooting up near by should be found numerous specimens of one of our small-flowered orchids, the ladies' tresses (*Spiranthes cernua*, Richard), with its delicate, white, sweet-scented flowers arranged in three ranks and crowded in a close spike. In similar places, here and there, a solitary small purple-fringed orchis (*Habenaria psycodes*, Gray) may be found. This is one of the showy orchids and is distinguished easily from the larger one by its size and later date of flowering. Under the shadow of the pines is the habitat of the two rattlesnake plantains (*Goodyera pubescens*, R. Br. and *G. repens*, R. Br.), with small, white, sac-shaped flowers crowded together on a stem six to twelve inches high, at the base of which is a thick cluster of small green leaves strongly marked with interlacing white lines. There, too, the coral-root (*Corallorhiza multiflora*, Nutt.) dwells, more solitary in its habit, but attracting attention by its strange brownish or yellowish color and the absence of green foliage. The flowers are dull-colored, and from ten to thirty in number, and the root is much branched and toothed, resembling a mass of coral, whence its name. Its appearance suggests to us two other plants found in similar situations, the Indian pipe (*Monotropa uniflora*, L.) and the pine-sap (*M. Hypopitys*, L.), which at a hasty glance might be mistaken for fungi, although they belong to the same family as the blueberry and the mountain laurel.

Another leafless parasitic plant, one of the most abundant in August, is the dodder (*Cuscuta Gronovii*, Willd.). It is found in low lands, often by the side of the railroads in such quantities as to attract the attention of the traveler despite the rapid rate at which he is whirling along. Its long, yellow, thread-like stems climb indifferently over herbs and shrubs. It differs from the parasites already mentioned by being parasitic above ground rather than below. In the early part of the summer its seed germinates in the ground, sending up a slender, twining stem. As soon as this reaches the surrounding herbage it attaches itself by suckers to the surface of the supporting plant. These suckers penetrate the outer part of the bark and absorb nourishment therefrom, while the root and the lower part of the stem wither away, leaving no connection with the ground, and causing the dodder to cling for dear life to the plant it has embraced. At first it is a very lowly plant;

> "But 'tis a common proof,
> That lowliness is young ambition's ladder,
> Whereto the climber-upward turns his face;
> But when he once attains the upmost round,
> He then unto the ladder turns his back,
> Looks in the clouds, scorning the base degrees
> By which he did ascend."

We can pause for only a moment to admire the purple violet-scented clusters of the ground-nut (*Apios tuberosa*, Moench.), and the nodding racemes of the hog peanut (*Amphicarpæa monoica*, Nutt.), but we must gather a handful of this charming white grass-of-Parnassus (*Parnassia Caroliniana*, Michx.), and of this curious-shaped turtle-head (*Chelone glabra*, L.). We must have some of these bronzy-blue gentians (*Gentiana*

*Andrewsii*, Griseb.) and some of these slender, humble hedge-hyssops (*Gratiola aurea*, Muhl.), with half a dozen or more ferns, including the dainty maidenhair (*Adiantum pedatum*, L.), our one New England representative of a large tropical genus of more than sixty species, many of which are among the choice treasures of the horticulturist, and some of the spleenworts, especially the delicate *Asplenium Trichomanes*, L., and *Asplenium ebeneum*, Ait., two of my favorites among the native ferns. August is so prolific in flowers that time would fail us to tell at large of the hawkweeds, the sunflowers, the St. John's-worts, the willow-herbs, the bur-marigolds and others, many of which bloom unseen and can endure to live unsung.

And the common wayside weeds! And the wild grasses! We must pass them by now unheeded, for we see by this little brook the queen of the August flowers, the bright, the beautiful, the far-seen, the cardinal flower (*Lobelia cardinalis*, L.). T. W. Higginson in his "Out-door Papers" says of it: "The cardinal flower is best seen by itself and, indeed, needs the surroundings of its native haunts to display its fullest beauty. Its favorite abode is along the dank mossy stones of some black and winding brook, shaded with overarching bushes, and running one long stream of scarlet with these superb occupants. It seems amazing how anything so brilliant can mature in such a darkness.

When a ray of sunlight strays in upon it, the wondrous creature seems to hover on the stalk, ready to take flight, like some lost tropic bird. There is a spot whence I have in ten minutes brought away as many as I could hold in both arms, some bearing fifty blossoms on a single stalk; and I could not believe that there was such another mass of color in the world." It fills in the August landscape the place of the rhodora in May, the mountain laurel in June and the water lily in July. Among the pictures that hang on the walls of my memory are some of a little pool, around the margin of which was a thick fringe of the cardinal flower, while in the waters the bright reflection seemed to double its beauty. When the meadows have been deprived of their wealth of grasses, and the nymphs of the brooks have hidden their faces from the heat of the August sun, and the distant hills are clothed with a hazy light, then we look for its coming and not in vain.

# SEPTEMBER FRUITS.

> Season of mists and mellow fruitfulness!
>   Close bosom-friend of the maturing sun;
> Conspiring with him how to load and bless
>   With fruit the vines that round the thatch-eaves run;
> To bend with apples the moss'd cottage-trees,
>   And fill all fruit with ripeness to the core;
> To swell the gourd, and plump the hazel shells
>   With a sweet kernel; to set budding more,
> And still more, later flowers for the bees,
>   Until they think warm days will never cease,
> For Summer has o'er-brimmed their clammy cells.
> 
> — KEATS— *To Autumn.*

All the rare days of the year are not confined to June. Although that is the time of the summer solstice, when the tide of life is nearing its flood, September has one of the high days of the year, the autumnal equinox, the day when, by Nature's time-table, the northern hemisphere is, like a train of cars, set off on a side track to wait while the sun moves southward to bring the life of the southern hemisphere from the siding where it has been waiting during its winter.

It is a time of fulness and content. Increase and multiply is Nature's motto. She has no sympathy with the slothful servant whose talent is hidden in a napkin. She believes in usury, in large per cents. To her a hundred per cent seems small; such per cents as two thousand, five thousand, ten thousand are more satisfactory. This September month is the time when she is busy paying her dividends; stock dividends they are, too, from her accumulated surplus of the year.

The time of fruits is at its prime, a happy season, sung by poets and praised by prose writers. In his seventh idyl Theocritus, writing more than two thousand years ago, describes a harvest feast on the island of Cos, east of the fair Ægean sea. It is an idyllic picture: "There we reclined on deep beds of fragrant lentisk, lowly strown, and rejoicing we lay in new stript leaves of the vine. And high above our heads waved many a poplar, many an elm tree, while close at hand the sacred water from the nymphs' own cave welled forth with murmurs musical. On shadowy boughs the burnt cicalas kept their chattering toil, far off the little owl cried in the thick thorn brake, the larks and finches were singing, the ring-dove moaned, the yellow bees were flitting about the springs. All breathed the scent of the opulent summer, of the season of fruits; pears at our feet and apples by our side were rolling plentiful, the tender branches with wild plums laden were earthward bowed, and the four-year-old pitch seal was loosened from the mouth of the wine-jars." And many a poet since Theocritus has touched upon the same theme.

On one of these fair September afternoons without a cloud, yet windless, with a gray haze shrouding the bright blue, neither burning overmuch, nor chill, let us yield to the instinctive love which urges us to leave behind for a little

"The vain low strife
That makes men mad — the tug for wealth and power —
The passions and the cares that wither life,
And waste its little hour;"

and while no promise of the fruitful year seems unfulfilled in this fair autumn tide, let us

"Roam the woods that crown
The uplands, where the mingled splendors glow,
Where the gay company of trees look down
On the green fields below."

But as one must come with an appetite to a feast in order to get the greatest enjoyment from it, so one must come to scenes of natural beauty with some taste for them or appreciation of them in order to enjoy them fully; for such, the September fields and woods have a charm of their own. All Nature seems to be pervaded by an air of rest and quiet. In the springtime we are drawn through glade and mead to enjoy the pleasure of seeing dormant life awaken, of watching the development of leaf-bud and blossom. They then seem so eager to unfold their beauty to us that the flowers can scarcely wait for the leaves to open before they spread their delicate petals to the wooing sun and the gentle breeze. The procession is to be a long one, and the earliest ones to appear are urged on, as it were, by an instinctive feeling that they must not delay nor encroach upon the time or space of those which follow. Like a well-managed railroad, Nature reserves the right to vary from her regular time-table as circumstances may require, but her trains usually make close connections. The flower of brief duration will find the short-lived insect at hand, whose aid is essential to its proper fertilization; the caterpillars will be snugly housed in their cocoons before the killing frosts come; the migratory birds take warning and keep near the sun; the winter store of grains or nuts is laid up in the cells of the little creatures which need them; winter with its needs is always preceded by the abundance of autumn.

There is no hurry about the later flowers. Few of them have the delicacy of the early ones. It is sufficient if there is time to ripen their fruits before winter.

That is the main object of their existence. Every part of their structure looks forward to the ripening of the fruit. And what a manifold variety of forms the fruit exhibits! Here, as elsewhere, it is only the attentive study which reveals all the truth.

Technically, the fruit consists of the matured pistil or gynæcium, including whatever may be joined with it. It is not necessarily edible, as fruit, in common language, is supposed to be. The word, fruit, is rather a loose term applied to a matured ovary, to a cluster of such ovaries when coherent, and to a matured ovary with the calyx and other parts of the flower attached to it. Fruits, therefore, are simple or complex, in various degrees, and have names to correspond. Simple fruits exhibit a great variety of forms, to which a corresponding variety of technical names has been applied. One is bewildered by such a list of terms as follicle, legume, loment, pyxis, silique, silicle, schizocarp, samara, caryopsis, utricle, akene and others, not to include the familiar nut, drupe, pome and berry, until a careful study of the forms of fruits has shown the characteristic differences and the necessity for some scheme of classification.

"By their fruits ye shall know them" is a good rule in botany and horticulture as in morals. Acorns grow on oaks, although the leaves may resemble those of the chestnut so strongly that the tree may be called the chestnut-oak. Grapes do not grow on thorns nor figs on thistles. In some families the fruit in all the species has the same general character and gives the name to the family. The fruit of the bean and the pea is a legume, and the great family, Leguminosæ, to which they belong has this distinguishing mark, that the fruit is always a legume or pod, or a simple modification of it known as a loment. In Ralph's "Icones Carpologicæ" there are figures of four hundred species from two hundred genera of this one family, and they form an interesting subject for study. Pods of all shapes and of great diversity in size, from that of the tiny clover to that of *Cassia Braziliana*, Lam., an inch and three-quarters in diameter and a foot and a half long, are faithfully represented.

In other families there is a great difference in the kinds of fruit in the different genera. In the Rose family we have drupes or stone-fruits in the cherry, peach and plum; pomes in the apple, pear and quince; follicles, pods which open on one side only, in the meadow-sweet and its kindred; and akenes variously arranged in the rose, the strawberry, the blackberry, the avens and cinquefoil.

The list of our September fruits, edible and inedible, palatable and unpalatable, is a long one, much longer in fact than anyone would suspect who had not made a collection of them. The profusion of bright-colored berries, where a little while before were summer flowers, is another of the charms of the month. Some of the most valued wild fruits are gathered before September. The strawberry, the three or four kinds of blueberries, the raspberries and some of the blackberries have already disappeared. But here in the huckleberry pasture, where the pyrolas and arethusa and quaking-grass and meadow-rue, loosestrife, roses, meadow-sweet, hardhack and the red summer-lily bloomed, the long-stemmed dangleberry (*Gaylussacia frondosa*, Torr. and Gray) is now in perfection. The pithy stems of the elder (*Sambucus Canadensis*, L.) are bending under their load of black-purple berries, which the catbird and others are doing their best to lighten. The high blackberry, in the sheltered places where it has not been parched by the August sun, yields its large juicy berries in great abundance. Perhaps we can, after a similar experience, appreciate Emerson's feeling toward them as expressed in his little poem :

> " 'May be true what I had heard,—
> Earth's a howling wilderness,
> Truculent with fraud and force,'
> Said I, strolling through the pastures,
> And along the river-side.
> Caught among the blackberry vines,
> Feeding on the Ethiops sweet,
> Pleasant fancies overtook me.

> I said, 'What influence me preferred,
> Elect, to dreams thus beautiful?'
> The vines replied, 'And didst thou deem
> No wisdom from our berries went?'"

The small cranberry (*Vaccinium Oxycoccus*, L.) may be of comparatively small value, but the large cranberry (*V. macrocarpon*, Ait.) is probably the most important of our native berries reduced to cultivation. The cranberry bog, like the huckleberry pasture, has become a considerable source of income when properly managed. The fruit of the cranberry tree (*Viburnum Opulus*, L.) is sometimes used as a substitute for the cranberry, which it much resembles in appearance. This shrub has two characteristics desirable in an ornamental shrub or tree: it is showy in flower and in fruit.

The wild black cherry and the choke-cherry display their long bunches of ripened fruit at this season, and the wild grapes are hanging over the walls or depending from the topmost branches of trees. The red berries of the barberry attract the eye as they hang like gems from every part of the bush. Under foot the red berries of the bearberry and the checkerberry are sought. Most of these, when not collected by man, furnish an important part of the food of birds and other wild creatures which depend on the bounty of Nature.

But the fruits which may be considered as inedible surpass these in number. The black berries of the withe-rod and the sweet viburnum, the blue of the arrow-wood and the purple of the maple-leaved viburnum, the pale blue berries of the silky cornel, the pale white of the panicled cornel, the bright red of the bunchberry and the flowering dogwood make these shrubs very attractive when in fruit, as they certainly are when in flower. Overtopping the low stone walls of the pastures are the sessile clusters of bright red berries of the black alder (*Ilex verticillata*, Gray), and in the swamps those of the smooth winterberry (*I. lævigata*, Gray) are far-seen, and the spice bush is gay with berries of the same color. The spikenard (*Aralia racemosa*, L.) and the wild elder (*A. hispida*, Vent.) are rich with dark-colored fruit, while the rarer ginseng (*A. quinquefolia*, Decsne. and Planch.) attracts us by its bright red cluster, and thus reveals its hiding-place. The odd-looking flowers of the Indian cucumber-root are now replaced by three or four black berries standing erect on the short stems which seemed unable to support the weight of the flowers, and the Solomon's seal bears now its dark-blue fruit, more attractive than its small greenish flowers. The tall stems of the pigeon-berry (*Phytolacca decandra*, L.) are loaded down with the long racemes of dark-purple berries filled with crimson juice. The corymbed clusters of bright red berries of the mountain ash (*Pyrus Americana*, DC.) have made it a favorite ornamental tree, and the purplish fruits of the choke-berry, abundant in pastures, are sometimes gathered carelessly with the huckleberries which they slightly resemble. The two baneberries (*Actæa alba*, Bigel. and *A. spicata*, L., var. *rubra*, Ait.) sometimes persist until September and are certain to be noticed if at all abundant. On the floor of the moist woodlands lie the creeping vines of the partridge-berry, of which the scarlet drupe crowned with the calyx teeth of two flowers forms a good illustration of a multiple or collective fruit.

These are not all the fruits which September yields, but they are enough to show the wealth which crowns this part of the year. The nuts of the hickories, hazels, chestnut, butternut and beech are now nearly ripe, but need the keen touch of frost before they reach their prime. The arrangement by which the fall of leaves from their branches, of acorns from their cups, of fruits of all sorts is brought about, is one which may well set the thoughtful mind on fruitful inquiry. When Sir Isaac Newton, sitting under that now famous apple-tree, asked why an apple falls to the ground, and answered his own question by saying that the earth attracts it, he was looking at the question from an entirely different point from that taken by the botanist, who would say that the fall of the apple depends on the presence and action of certain cells at the base of the stem, by an arrangement provided for even in the bud. The scar left where a leaf-stem or a fruit-stem has separated from the branch indicates in some degree how this is brought about.

September has, like August, some of the bright-colored flowers. The golden-rods and asters which appeared in August are reënforced by others during this month. It is not difficult now to find a dozen species of aster, some of which as *Aster vimineus*, Lam. and *A. multiflorus*, Ait. are very common. In fact, if we except the genus *Carex*, the genus *Aster* is our most abundant genus so far as the number of species, if not of individuals, is concerned. On the gravelly margin of the ponds, between high and low water mark, the little pipewort still rears its dense heads of small flowers, and close by it the little hedge hyssop gives a tinge of yellow to the shore. In some little pools the trailing stems of the water purslane (*Ludwigia palustris*, Ell.) suggest the almost ineradicable purslane of the gardens. Coarse grasses are now in the forefront. Barnyard-grass (*Panicum Crus-galli*, L.) and old-witch grass (*P. capillare*, L.) and crab-grass (*P. sanguinale*, L.), the foxtails (*Setaria glauca*, Beauv. and *S. viridis*, Beauv.), the beard-grasses (*Andropogon scoparius*, Michx. and *A. furcatus*, Muhl.) by the wayside, and the wood-grass (*Chrysopogon nutans*, Benth.) and the drop-seed grass (*Muhlenbergia Willdenovii*, Trin.) and the wood reed-

## Swamp Meadow.

*"I suppose that these meadows are as nearly in their primitive state as any that we see. So this country looked, in one of its aspects, a thousand years ago.'*

—H. D. THOREAU.

grass (*Cinna arundinacea*, L.) of the woodland add to the luxuriance of vegetation which is one of the great charms of untrimmed and unpruned Nature.

> "I know the lands are lit
> With all the autumn blaze of Golden Rod;
> And everywhere the Purple Asters nod
> And bend and wave and flit,"

but September will hardly seem to be September if before its last days I do not somewhere find the favorite among the late flowers, the fringed gentian (*Gentiana crinita*, Froel.). To see it growing in its native haunts in some open glade of the forest, or by the margin of some dusky spring or in the lowlands by the roadside, with numerous blossoms opening their sweet and quiet eyes to the sky is one of the compensations of autumn. Bryant's little poem, "To the Fringed Gentian," is probably the most familiar of all the American poems of that kind. It is one of a group including, among many others, "To a Waterfowl," "Autumn Woods," "The Painted Cup," "The Planting of the Apple-tree," "Robert of Lincoln," which show his deep poetic insight into Nature. He too, like Wordsworth, could say:

> "To me the meanest flower that blows can give
> Thoughts that do often lie too deep for tears,"

but the whole world is richer for the poetic expression of the thoughts for which he found words. In the interval between 1812 when he wrote "Thanatopsis" and 1876 when he wrote its counterpart, "The Flood of Years," there is no decay of the high poetic faculty, ever the same serene song. It is not easy to estimate exactly the influence of such poetry upon the life of a people, but it is very plain that, if we had more little poems like "To the Fringed Gentian" and Emerson's "Rhodora," Wordsworth's "Daffodils" and Burns' "To a Mountain Daisy," even with the greatly changed conditions of modern life, our flora would assume a greater interest in our minds, and the pleasure of living would have an added zest.

After the fringed gentian has bloomed we know there is nothing new to be expected save the fantastic witch-hazel. Like the witch-elm of Great Britain, it was formerly used for divining rods, and I have seen it used in recent years to determine the location of water before the digging of a well. Its magic powers might have been originally suggested by its peculiar habit of bearing flowers in the autumn even after its leaves have fallen, thus reversing the general order of Nature; perhaps, by the fact that it does not ripen its fruit-capsules until the following summer, so that buds, flowers and fruit may be found in perfection upon it at the same time. It is easily identified by its clusters of four or five yellow flowers in the axils of the leaves, with their long linear petals. When its blossoms have faded, we must wait until a new year, wait in hope for the resurrection of life which is sure to come with ever-renewed beauty.

# THE RECORD OF ONE YEAR

WITH

ADDITIONS FROM OTHER YEARS.

## NOTE.

FOR several years prior to 1882, and later, I kept a record of the dates on which I saw for the first time in each year any wild flower or fern. In the eleven papers which follow I have given the record for 1882, dividing the plants into groups of fifty each, numbered with Arabic figures. I have annexed for each month except April a supplementary list, numbered in Roman notation, found during the same month in other years, but not seen in 1882. I think this is a fairly representative list of the number which may be found by one person in any town in Worcester County. More than ninety-five per cent of these were found in Millbury. There will naturally be variations in such lists, and this is not offered as being complete, but suggestive. The nomenclature has been conformed to that of the sixth edition of Gray's Manual, as the book most widely used for reference in this subject in New England. I am aware that many changes in nomenclature have already been proposed and that others are to follow, but it has not seemed best to give names which, as yet, are to be found only in special works. When the nomenclature of our flora has been settled on a satisfactory and permanent basis, perhaps there may be an opportunity for the revision of this little book, in which such desirable changes might be properly made.

# THE EARLIEST FLOWERS.

> The first conscious thought about wild flowers was to find out their names — the first conscious pleasure, — and then I began to see so many that I had not previously noticed. Once you wish to identify them there is nothing escapes, down to the little white chickweed of the path and the moss of the wall. . . . The instant you look for them they multiply a hundred fold; if you sit on the beach and begin to count the pebbles by you, their number instantly increases to infinity by virtue of that conscious act.
>
> — RICHARD JEFFERIES — *The Open Air.*

THERE are many gates that lead into the countless fields of the Earthly Paradise. I remember as if it were yesterday, when I found a key that opened one of them. The date is there in my diary, May 10; the year ——, well, it is not so long ago.

The young lad wanders through the fields with his comrades and learns from them the names of a few living things that attract their attention, and the childish legends filling him with wonder that are part of the inheritance of the race. The world is beautiful, the glory and the freshness of a dream of youth are over it all. He does not trouble himself to analyze the causes of his joy in beholding it. It is enough for him that it is so fair and pleasant. Why, after all, should he trouble himself to find out how the apple got inside the dumpling? That question, so far as he is concerned, may safely be left to the king.

And all the time, his memory is unconsciously preserving these things, piled as it were one upon the other, forming palimpsests that will be the cherished treasures of later years. But there comes a time when, perhaps, he desires to know more of the world around him and to know it in a different way, and then the key to one of the gates is in his hand. Desire to know is the "Open Sesame" at which the gate will turn on its hinges. Some one may have entered at the same gate, have traveled far afield and on his return have told the story of the sights to be seen in that pleasant land; and the reading of the little tale has been an inspiration to many another to follow in the same path.

The dog's tooth violet (*Erythronium Americanum*, Ker.) was the first flower which I was able to identify by a careful study of its structure. I must have seen fields yellow with it often before, but I cannot think that I ever really noticed it till then. It certainly seems different ever since. I look for it every spring time; if I should not see it, something would be wanting to the perfect charm of that delightful season. The next date in that early diary is May 14th, and the next names in the list are wild oats (*Oakesia sessilifolia*, Watson) and fringed polygala and wake robin and anemone; and the list goes on lengthening as the spring fades into summer and summer declines into autumn, until with the blooming of the witch hazel in the early October days I found that the delightful labor of that year was ended. There are two hundred and thirty-one names in the list. I had found not only these, most of which had been hitherto unnoticed, but with them a keener interest in the unrealized beauties of the world lying close to my feet and a deeper sense of the all-embracing mystery which surrounds our life.

The next year I went over the same road and the list grew to three hundred and seventy-six, and with each succeeding year new names have been added till now it would not be difficult to find six hundred of them in the narrow round in which one can travel within three or four miles of home. When, years afterward, I read Jefferies' "Field and Hedgerow," and came upon that touching passage written in his fatal sickness, I think I could realize his yearning, by having a similar love. It is in the first paper, "Hours of Spring," beautiful and yet so pathetic.

"I wonder to myself how they can all get on without me — how they manage, bird and flower, without me to keep the calendar for them. For I noted it so carefully and lovingly, day by day, the seed-leaves on the mounds in the sheltered places that came so early, the pushing up of the young grass, the succulent dandelion, the coltsfoot on the heavy thick clods, the trodden chickweed despised at the foot of the gate-post, so common and small, and yet so dear to me. Every blade of grass was mine, as though I had planted it separately. They were all my pets, as the roses the lover of his garden tends so faithfully." So true is it that the further one goes inside the gate the wider the fields stretch away to the infinite spaces.

In these early spring days when the buds of the elms that line the village streets are swelling nigh to bursting into purple fascicles, and the catkins of the willows not yet fully developed are sought for their promise of spring, in the swamp by the riverside where alders and red maples and poison dogwood grow thick, I expect to find the skunk-cabbage in bloom. The name and—the pity of it—the odor are enough to ruin the reputation of a far fairer flower. It is the herald of the long procession into which its glory fades as that of the morning star into the sun. The purple and green spathes scattered about tell me that I have not come too soon. On drawing aside the hoods of several of them I find the flowers are faded and the fruit is ripening. But there are a few just peering from the ground, and these are in their flowering prime and, therefore, they shall stand at the head of the list for this year.

## THE EARLIEST FLOWERS.

By the foot-path that leads over the hill I find the beaked hazel (*Corylus rostrata*, Ait.) in bloom. The little red stigmas of the fertile flowers are scarcely thrust forth from the scaly bud and the crowded stamens of the sterile flowers are peeping out from the little roofs that cover them. The wind will have to be the kindly messenger to carry the pollen whither it should go. I am not alone on my ramble.

> "I hear from many a little throat
>   A warble interrupted long;
> I hear the robin's flute-like note,
>   The bluebird's slenderer song.
> Brown meadows and the russet hill,
>   Not yet the haunt of grazing herds,
> And thickets by the glimmering rill
>   Are all alive with birds."

A flock of robins goes trooping before me among the apple-trees, and a pair of red-winged blackbirds fly startled away. The scream of a bluejay is heard from that clump of trees on the right, and in the woods all around the crows are apparently holding a caucus. The song sparrow greets me from many a tree by the wayside and his brother, "Little Chippy," shows his brown pate more than once. A flock of wild ducks, black-headed and white-bodied at this distance, is floating lazily on the surface of the pond yonder.

While the conspicuous flowers have not yet made their appearance, the less known and more humble mosses may claim attention. These are to be found almost everywhere, on the stones by the roadside, as soft cushions in the shade of the forest, clothing the base of trees in the lowlands, and forming the greater part of the peat swamps. Their structure is so minute that it cannot be studied without a microscope, and yields in fascination to no branch of botanical study.

The lichens, too, are now abundant. They differ from the mosses in that their nutrition is derived from the atmosphere. It is found that they will not grow in a flourishing condition in the immediate neighborhood of large cities where the air is filled with smoke or deleterious gases; so that their presence in a fully developed condition is an indication of the purity of the atmosphere. Lichens, unlike mosses, have been applied to many uses, but their use is naturally diminishing as civilization and, especially, the science of chemistry have advanced. They are of slow and, frequently, of very long growth. They are of various colors, green, white, yellow, brown and black being very common. Some hang from the stems and limbs of trees in threadlike forms, and produce the shaggy and venerable appearance characteristic of the aged tree in the forest. Some cling so closely to the surface of the rocks that the severest storms cannot detach them. They spend their lives in preparing places for the higher forms which come after them. The study of them will enable us to appreciate Ruskin's eloquent praise of them:

"Meek creatures! the first mercy of the earth, veiling with hushed softness its dentless rocks, creatures full of pity, covering with strange and tender honor the sacred disguise of ruin, laying quiet fingers on the trembling stones to teach them rest. No words that I know of will say what these mosses and lichens are; none are delicate enough; none perfect enough; none rich enough. They will not be gathered like the flowers for chaplet or love-token, but of these the wild-bird will make its nest and the wearied child its pillow, and as the earth's first mercy, so they are its last gift to us. When all other service is vain from plant and tree, the soft mosses and grey lichens take up their watch by the headstone. The woods, the blossoms, the gift-bearing grasses, have done their parts for a time, but these do service forever. Trees for the builder's yard, flowers for the bride's chamber, corn for the granary, mosses and lichens for the grave."

While we have been turning aside, the warm April sun has been allowing other blossoms to unfold. It makes real the myth of the "Fountain of Youth"; but Nature does not disclose all her secrets even to wishful eyes, and all the charms of this upspringing of life in its wonderful variety are not yielded even to earnest and persevering search. And yet, as

> "'T is better to have loved and lost
> Than never to have loved at all,"

so it is better to have made the search, even if all the treasure could not be found. Treasures of an unexpected kind are thus often found, sometimes of more value than those that were sought. A firmer health, a fuller mind, a deeper feeling of content, a broader view of life, are not these worth finding? Are they not worthy of long and anxious seeking?

But it is time to introduce the first list of fifty. For more than one reason it is worth looking at.

| April 3. | 1 | Symplocarpus fœtidus, Salisb. | Skunk Cabbage. |
| | | Corylus rostrata, Ait. | Beaked Hazel. |
| " 8. | | Corylus Americana, Walt. | Common Hazel. |
| | | Epigæa repens, L. | Trailing Arbutus. |
| " 14. | | Ulmus Americana, L. | American Elm. |
| " 16. | | Populus tremuloides, Michx. | American Poplar. |
| | | Alnus incana, Willd. | Speckled Alder. |
| " 20. | | Acer rubrum, L. | Red Maple. |
| | | Houstonia cærulea, L. | Innocence. |
| | 10 | Antennaria plantaginifolia, Hook. | Everlasting. |
| | | Sanguinaria Canadensis, L. | Bloodroot. |
| | | Hepatica triloba, Chaix. | Hepatica. |
| | | Salix discolor, Muhl. | Willow. |

## THE EARLIEST FLOWERS.

| | | | |
|---|---|---|---|
| April | 21. | Carex Pennsylvanica, Lam. | Sedge. |
| | | Myrica Gale, L. | Sweet Gale. |
| " | 22. | Saxifraga Virginiensis, Michx. | Early Saxifrage. |
| | | Anemone nemorosa, L. | Anemone. |
| | | Lindera Benzoin, Blume | Spice-bush. |
| " | 24. | Ulmus fulva, Michx. | Slippery Elm. |
| | | 20 Viola rotundifolia, Michx. | Round-leaved Violet. |
| | | Dirca palustris, L. | Leatherwood. |
| " | 26. | Stellaria media, Smith | Chickweed. |
| " | 28. | Erythronium Americanum, Ker | Dog's-tooth Violet. |
| " | 29. | Populus grandidentata, Michx. | Large-toothed Poplar. |
| | | Viola sagittata, Ait. | Arrow-leaved Violet. |
| May | 1. | Erodium cicutarium, L'Her. | Stork's-bill. |
| | | Chrysosplenium Americanum, Schwein. | Golden Saxifrage. |
| " | 2. | Taraxacum officinale, Weber | Dandelion. |
| | | Salix humilis, Marsh. | Prairie Willow. |
| " | 3. | 30 Viola canina, L., var. Muhlenbergii, Gray | Dog Violet. |
| | | Capsella Bursa-pastoris, Moench | Shepherd's Purse. |
| | | Fragaria Virginiana, Mill. | Strawberry. |
| | | Cerastium vulgatum, L. | Mouse-ear Chickweed. |
| " | 4. | Caltha palustris, L. | Marsh Marigold. |
| | | Carex umbellata, Schkuhr | Sedge. |
| | | Myrica asplenifolia, Endl. | Sweet Fern. |
| | | Potentilla Canadensis, L. | Cinque-foil. |

May 4.   Luzula campestris, DC.            Wood Rush.
         Viola blanda, Willd.              Sweet White Violet.
      40 Aralia trifolia, Decsne. & Planch.  Ground-nut.
         Anemonella thalictroides, Spach   Rue Anemone.
"    5.  Viola palmata, L., var. cucullata, Gray   Blue
           Violet.
         Trillium erectum, L.              Purple Trillium.
         Oakesia sessilifolia, Watson      Wild Oats.
"    6.  Cassandra calyculata, Don         Leather-leaf.
         Arctostaphylos Uva-ursi, Spreng.  Bearberry.
         Viola pedata, L.                  Bird-foot Violet.
"    8.  Caulophyllum thalictroides, Michx. Pappoose-
           root.
         Ranunculus fascicularis, Muhl. Early Buttercup.
      50 Asarum Canadense, L.              Wild Ginger.

In 1881 the record shows almost the same list and exactly the same number on the 8th of May, so that I think it may be taken as a fairly representative list of this immediate neighborhood. There are omissions, of course. It is a truly democratic collection; the lordly tree, the humble weed, the common and the rare stand here side by side.

Sixteen of them, nearly a third of the whole, are trees or shrubs, six are violets, the genus most largely represented, and the others are divided among various families and genera.

There are favorite flowers among them. One of these is the round-leaved violet, which Bryant must have found as he roamed in the woods about Cummington brooding over "Thanatopsis," and of which he wrote:

"When beechen buds begin to swell,
    And woods the bluebird's warble know,
The yellow violet's modest bell
    Peeps from the last year's leaves below.

Ere russet fields their green resume,
    Sweet flower, I love, in forest bare,
To meet thee, when thy faint perfume
    Alone is in the virgin air.

. . . . . . .

Oft in the sunless April day,
    Thy early smile has stayed my walk;
But midst the gorgeous blooms of May,
    I passed thee on thy humble stalk.

So they, who climb to wealth, forget
    The friends in darker fortunes tried.
I copied them — but I regret
    That I should ape the ways of pride."

In Sir J. D. Hooker's "Himalayan Journals" there is a reference to one of the humbler plants in this list which is well worth quoting as one of the best expressions of the delights of scientific travel. He is writing under date of November 25th, 1848, on one of the passes of the Himalayas in East Nepaul, at an elevation of 13,000 feet. The grass which he mentions (*Poa annua*, L.) is No. 55 of my next list. He says:

"Along the narrow path I found the two commonest of all British weeds, a grass (*Poa annua*), and the shepherd's purse! They had evidently been imported by man and yaks, and as they do not occur in India, I could not but regard these little wanderers from the north with the deepest interest. Such incidents as these give rise to trains of reflection in the mind of the naturalist traveler; and the farther he may be from home and friends, the more wild and desolate the country he is exploring, the greater the difficulties and dangers under which he encounters these subjects of his earliest studies in science, so much keener is the delight with which he recognizes them, and the more lasting is the impression which they leave. At this moment these common weeds more vividly recall to me that wild scene than does all my journal, and remind me how I went on my way, taxing my memory for all it ever knew of the geographical distribution of the shepherd's purse, and musing on the probability of the plant having found its way thither over all Central Asia, and the ages that may have been occupied in its march."

Every one familiar with the native flora must have had similar, if not quite as striking, experiences. The old roads, the same flowers seen year after year, these are a perennial delight, and they open the heart for the new ones which we chance to meet afar.

# THE FLOWERS OF MAY.

> Now the bright morning-star, day's harbinger,
> Comes dancing from the east, and leads with her
> The flowery May, who from her green lap throws
> The yellow cowslip and the pale primrose.
>     Hail, bounteous May, that dost inspire
>     Mirth and youth and warm desire;
>     Woods and groves are of thy dressing;
>     Hill and dale doth boast thy blessing!
> Thus we salute thee with our early song,
> And welcome thee, and wish thee long.
>         — MILTON — *Song on May Morning.*

In the palmy days of Athens, Pindar could invent no prouder titles for that preëminent city, none which could give the Athenians greater pleasure, than these three, Fruitful, Famous in Story, Violet-crowned. The month of May, rich in promises, famous in song, is, most of all, the month of the violet crown. All our violets can be found in May, most of them in great abundance, so that they form one of the most characteristic groups of spring flowers.

About a hundred species of violets have been described, mostly from the North Temperate Zone. Two of them, not known to be natives of North America, have been much cultivated; the sweet violet (*Viola odorata*, L.) on account of its fragrance, and the pansy or heartsease (*V. tricolor*, L.) on account of its beautiful flowers. The pansy is one of the finest of the florists' flowers, and has been much improved by cultivation. The difference between some of its wild and some of its cultivated forms is so great that it is hard to realize that the latter have been evolved from the former. Its name is from the French *pensée*, suggestive of thoughtfulness, probably from the slightly drooping habit of the flower. It is to this that Shakspere alludes in Hamlet, when Ophelia says:

"There's rosemary, that's for remembrance; pray you, love, remember; and there is pansies, that's for thoughts."

"In Warwickshire and Worcestershire this plant is called by the common people *Love in Idleness*," says Thornton, "and therefore is doubtless the herb to which the inventive fancy of Shakspere attributes such extraordinary virtues in the person of Oberon, king of the fairies, in 'The Midsummer Night's Dream'":

"Yet mark'd I where the bolt of Cupid fell:
It fell upon a little western flower,
Before milk-white, now purple with love's wound,
And maidens call it love-in-idleness.

> Fetch me that flower; the herb I showed thee once:
> The juice of it on sleeping eyelids laid
> Will make or man or woman madly dote
> Upon the next live creature that it sees."

In another of his plays, notably the last complete one, "The Winter's Tale," the one in which the golden glow of the sunset of his genius tinges the russet mantle of the morn, the gentle Perdita leads us by the hand through the meadows by the side of Avon while she gathers in thought

> "Daffodils,
> That come before the swallow dares, and take
> The winds of March with beauty; violets dim,
> But sweeter than the lids of Juno's eyes
> Or Cytherea's breath; pale primroses,
> That die unmarried, ere they can behold
> Bright Phœbus in his strength; bold oxlip and
> The crown imperial."

Milton, too, is not behind Shakspere in his appreciation of these lovely flowers. When describing the bower of Adam and Eve in Paradise, he gives the violet a place:

> "Under-foot the violet,
> Crocus and hyacinth, with rich inlay
> Broider'd the ground, more colour'd than with stone
> Of costliest emblem:"

and again:

> "Flowers were the couch,
> Pansies and violets and asphodel
> And hyacinth; earth's freshest, softest lap."

In his "Lycidas" he gives a short list of the vernal flowers that purple all the ground:

> "Bring the rathe primrose that forsaken dies,
>   The tufted crow-toe and pale jessamine,
> The white pink, and the pansy freak'd with jet,
> The glowing violet,
> The musk-rose and the well-attired woodbine,
> With cowslips wan that hang the pensive head."

In "The Pilgrim's Progress" as Christiana and her children are going through the Valley of Humiliation with Mr. Great-heart, they hear a shepherd's boy singing:

> "He that is down needs fear no fall;
>   He that is low, no pride;"

and the guide says, "Do you hear him? I will dare to say that this boy lives a merrier life, and wears more of that herb called heart's-ease in his bosom, than he that is clad in silk and velvet."

In Britten and Holland's "Dictionary of English Plant-Names," there is a list of forty-one common names of the pansy, used in the different counties of England.

Some of them are very quaint. One of these, "Meet-her-i'-th'-entry-kiss-her-i'-th'-buttery," is probably the longest plant-name in the English language, but is only one of a great number of similar names that have been given to this flower. This particular one is local in northwestern Lincolnshire. The progress of modern education and the increased facilities of communication will soon make such names only souvenirs of a simple and homely Past.

The violet has long been in popular language the emblem of modesty. It is born in the grass in an humble and obscure situation, it is true; it does not conceal itself there. It reaches up into the light, and both by its color and by its perfume attracts attention. It became the imperial flower of France, closely associated with the Napoleonic dynasty, as the fleur-de-lis is associated with the royal line. Its color has become associated with high rank. It has covered the heads of the Church, the archbishops and the bishops, with its livery; it is the color of the mourning of kings. It was one of the favorite colors of the old poets, especially as applied to the sea. Matthew Arnold has caught the spirit of the Greek muse when, referring to the Gods sitting on high on Olympus, he says:

"The Gods are happy.
They turn on all sides
Their shining eyes,
And see below them
The earth and men.

They see the Heroes
Sitting in the dark ship
On the foamless, long-heaving
Violet sea,
At sunset nearing
The Happy Islands."

Violet is the most refrangible of the colors of the spectrum. According to the undulatory theory of light, the number of vibrations of the æther necessary to produce violet color is about 760 million million per second, and the wave-length in air about sixteen millionths of an inch. It is the reflection of such vibrations or waves that enables us to see the violets and appreciate their color.

The sweet violet and the pansy, these two favorites of the genus, are so well known to us by cultivation that any allusion to them seems as familiar as if it referred to our native species; in many cases it is more familiar, owing to the fact that so little attention is paid to those which grow naturally and freely all about us. Yet our native species have qualities of their own which give them some claim upon us. Some of them are blue, some are white and some are yellow. All of them have exquisite pencillings of color which probably serve as guides to the insects which visit them. They have elaborate contrivances to secure cross-fertilization, contrivances so elaborate that we are led to believe that we are witnessing almost the maximum development of the insect-fertilizing idea which has been such a prominent feature of modern botanical thought. For, besides the conspicuous blossoms, which appear early, there are others produced later, which never open, in which petals are absent or rudimentary; but these are fertilized in the bud and are much more prolific than the ordinary flowers, though they are apt to escape observation from their manner of growth. As Massee puts it: "The whole structure of the (conspicuous) flower shows that self-fertilization was almost impossible, and that the visits of insects were indispensable, yet all these elaborate arrangements have not prevented the violets from evolving something even more effectual and at the same time more economical in connection with fertilization, and in reality the old-fashioned colored flower, evolved for aiding insect-fertilization, is now of no use to the plant. But, when a structure is evolved, it cannot be at once arrested, even when completely superseded and useless, and this is the condition that most of the violets now find themselves in, encumbered with an old effete type of arrangement for securing insect-fertilization. The new type of flower present in many violets appears later in the season than the old type of flower, from which it differs fundamentally in being self-fertilized, the sepals remaining closed until fertilization is effected. The reversion to the self-fertilized method seems, and really is, unintelligible from the present stand-point of knowledge, assuming that cross-fertilization really invigorates and helps plants in the struggle for existence; nevertheless, in the case of the violets and many other groups, this is certainly the direction in which things are tending at the present day."

Nine or ten species of wild violets can be found within a short radius in almost any part of our county. My favorites among them are the early yellow violet, which is rare further south except among the mountains, the downy yellow violet (*Viola pubescens*, Ait.), the sweet white violet (*V. blanda*, Willd.), and the birdfoot violet (*V. pedata*, L.). They all differ more or less in habit. Most of them are stemless, leaves and flowers springing from the root. Some are found in moist lowlands, others in dry upland pastures and thickets. Most of them are easily distinguished by their leaves. In the sandy soil of the railroad embankment, in such places where the gray birch loves to dwell, the thick rootstocks of the bird-foot violet send up numerous scapes tipped with light blue flowers. In my mind's eye now there is a picture of a field covered with these pretty flowers as thickly as an English or Scottish moorland with the purple heather, and leaving as attractive memories. It belongs now to the long ago, yet it helps me to appreciate Tennyson's feeling for the sweet violet:

> "The smell of violets hidden in the green
>     Poured back into my empty soul and frame
>   The times when I remembered to have been
>     Joyful and free from blame."

But while I have been tarrying with the violets the long procession of the vernal flowers has been moving forward. What a pleasure, to welcome them again on their brief annual visit! How unique is the appearance of the forest when putting on its spring dress! The pale green of the poplars now shows afar, the dark pines are losing that prominence which makes them so marked a feature of the winter woods, the variety of tints in the plaited foliage combines to present an appearance seen at no other time of the year, and seen now in its perfection only when illumined by the bright rays of the sun nearing the horizon. It is interesting to watch the characteristic tints of the spring foliage change under the bright sunlight of late May and June into the more nearly uniform color of midsummer, and so remain until the ripening days of autumn again bring out characteristic colors.

The feeling of overflowing life, the joy of mere existence, the sense of awakening to a new bright day full of hope seems to be realized in the flowers of May. It is a time of promises, some of which will surely be fulfilled; of hopes, some of which will surely be realized. It is well for us if we can open our hearts to these sweet influences of Nature, so that the beauty of flowers, the charm of birds, the glorious possibilities of Life may keep those hearts young, so that we can look on the face of the earth with all the wonder of youth, refined by the wisdom that comes with experience and years.

May 8. 51 Ranunculus abortivus, L. Small-flowered Buttercup.
Salix sericea, Marsh. Silky Willow.
Equisetum arvense, L. Common Horsetail.
" 9. Acer saccharinum, Wang. Sugar Maple.
Poa annua, L. Low Spear Grass.
" 10. Vaccinium vacillans, Solander Low Blueberry.
Aquilegia Canadensis, L. Wild Columbine.
Amelanchier Canadensis, Torr. & Gray Shadbush.
Viola pubescens, Ait. Downy Yellow Violet.
" 13. 60 Arisæma triphyllum, Torr. Jack-in-the-Pulpit.
Vaccinium corymbosum, L. Swamp Blueberry.
" 16. Vaccinium Pennsylvanicum, Lam. Dwarf Blueberry.
Veronica serpyllifolia, L. Thyme-leaved Speedwell.

*Rhododendron Rhodora*, Don.
The Rhodora.

"*In May when sea-winds pierced our solitudes*
*I found the fresh Rhodora in the woods.*"
—R. W. Emerson.

## THE FLOWERS OF MAY. 143

|  |  | Trillium erythrocarpum, Michx. | Painted Trillium. |
|---|---|---|---|
|  |  | Coptis trifolia, Salisb. | Three-leaved Goldthread. |
|  |  | Thalictrum dioicum, L. | Early Meadow Rue. |
|  |  | Ranunculus bulbosus, L. | Common Buttercup. |
|  |  | Rumex Acetosella, L. | Field Sorrel. |
|  |  | Thaspium aureum, Nutt. | Meadow Parsnip. |
| May 19. | 70 | Viola lanceolata, L. | Lance-leaved Violet. |
|  |  | Ribes oxyacanthoides, L. | Gooseberry. |
|  |  | Anthoxanthum odoratum, L. | Sweet Vernal Grass. |
|  |  | Carex bromoides, Schkuhr | Sedge. |
| " | 23. | Salix petiolaris, Smith | Willow. |
|  |  | Polygala paucifolia, Willd. | Fringed Polygala. |
|  |  | Trillium cernuum, L. | Wake Robin. |
|  |  | Streptopus roseus, Michx. | Twisted Stalk. |
|  |  | Viburnum lantanoides, Michx. | Hobble-bush. |
|  |  | Carpinus Caroliniana, Walter | American Hornbeam. |
|  | 80 | Ostrya Virginica, Willd. | American Hop-Hornbeam. |
|  |  | Rhododendron Rhodora, Don | Rhodora. |
|  |  | Sambucus racemosa, L. | Red-berried Elder. |
|  |  | Carex vestita, Willd. | Sedge. |
| " | 26. | Polygonatum biflorum, Ell. | Solomon's Seal. |
|  |  | Betula populifolia, Ait. | Gray Birch. |
|  |  | Erigeron bellidifolius, Muhl. | Robin's Plantain. |
|  |  | Carex laxiflora, Lam. | Sedge. |

|  |  | Uvularia perfoliata, L. | Bellwort. |
|---|---|---|---|
| May 27. |  | Salix lucida, Muhl. | Shining Willow. |
|  | 90 | Prunus Pennsylvanica, L. f. | Wild Red Cherry. |
|  |  | Saxifraga Pennsylvanica, L. | Swamp Saxifrage. |
|  |  | Euphorbia Cyparissias, L. | Spurge. |
|  |  | Chelidonium majus, L. | Celandine. |
|  |  | Carex stricta, Lam. | Sedge. |
| " | 29. | Quercus rubra, L. | Red Oak. |
|  |  | Fraxinus Americana, L. | White Ash. |
|  |  | Aralia nudicaulis, L. | Wild Sarsaparilla. |
|  |  | Geranium maculatum, L. | Wild Cranesbill. |
|  |  | Maianthemum Canadense, Desf. | False Solomon's Seal. |
|  | 100 | Pedicularis Canadensis, L. | Lousewort. |

There are seasons when some of the common flowers escape us. Catkins, except those of the willows, seem to be unusually scarce this year. Of the four birches found here, the yellow, the black, the paper and the gray, I have found those of the last only, and these in very small quantity, where in other years all of them were very abundant. The same may be said of some other trees. All of them are somewhat behind the usual time of flowering, so that the above dates can hardly be regarded as the average ones, but should be a week or ten days earlier.

It will be noticed that eighteen of the fifty are trees or shrubs. Among the rarer forms in this immediate neighborhood is the red-berried elder (*Sambucus racemosa*, L.), although in the rocky gorge of Purgatory, a dozen miles away, it is one of the most abundant shrubs. Another is the wild gooseberry (*Ribes oxyacanthoides*, L.), of which I can find but one or two shrubs. The American hornbeam (*Carpinus Caroliniana*, Walter) and the hop-hornbeam (*Ostrya Virginica*, Willd.) are small, handsome trees. The smooth, fluted trunk of the former is an interesting object to a person curious in forest history. Its fruit is unlike any other. Its foliage somewhat resembles that of a birch. The foliage of the hop-hornbeam also resembles that of the black birch, and its fruit has a resemblance to that of the hop. Neither of these is as well known as it deserves to be. The dark knobs that now stud the white ash disclose one of our most valuable forest trees.

These apparently withered stems in the grass by the roadside are the fertile stems of the common horsetail (*Equisetum arvense*, L.), the spores of which are interesting microscopic objects, illustrating finely the influence of moisture on the attached filaments. Goldthread and early meadow-rue and the sedges and bellwort and Solomon's seal and celandine and cranesbill, each and all seem to be needed to complete the charm of the month. If we look at them with appreciative eye, they will give us a keener enjoyment of Life.

# THE FLOWERS OF EARLY JUNE. I.

> O month whose promise and fulfilment blend
> And burst in one! it seems the earth can store
> In all her roomy house no treasure more;
> Of all her wealth no farthing have to spend
> On fruit, when once this stintless flowering end.
> And yet no tiniest flower shall fall before
> It hath made ready at its hidden core
> Its tithe of seed, which we may count and tend
> Till harvest.
> — HELEN JACKSON — *June*.

It is one of the rewards which impartial Nature bestows upon her faithful students that their eyes are opened to see the order, and their ears to hear the harmony, and their minds to perceive the beauty, which pervade the universe. To them alone "the music of the spheres" is audible; to them alone "the open secret" of Life is revealed. And yet, Nature, however kindly, does not bestow all her gifts upon any one of her followers however faithful. Often the light of new truth appears to dawn on several watchers at once, just as the early morning light dawns upon all who are awake to see it.

The independent discovery of the planet Neptune in 1846 by Adams and Leverrier by means of the disturbance it produces in the movements of Uranus will always be regarded as one of the most interesting events in the history of astronomy. Equal honors were justly awarded to both in 1848 by the council of the Royal Astronomical Society. Newton and Leibnitz were working simultaneously on mathematical theories which led to the invention of the differential calculus. We may now forget the long and bitter controversy which followed to determine to which of the two belongs the right of claiming priority in this invention.

While Darwin, by the advice of Lyell, was writing out his views on the tendency in organic beings descended from the same stock to diverge in character as they become modified, views afterwards embodied in his "The Origin of Species," Wallace, who was then in the Malay Archipelago making his famous researches, sent him an essay "On the tendency of Varieties to depart indefinitely from the Original Type," containing exactly the same views as his own. The modesty of both allowed no question of priority to disturb their friendship and respect for each other. The 14th of February, 1876, must be memorable in the records of the United States Patent Office as the day on which Elisha Gray filed a *caveat* in regard to his own invention, and Alexander

Graham Bell filed a specification and drawings of the original Bell telephone.

Professor Asa Gray, writing of what is generally called Metamorphosis of Plants, says: "The adopted theory supposes that stamens and pistils, as well as sepals and petals, are homologous with leaves; that the sepals are comparatively little, the petals more, and the reproductive organs much modified from the type, that is, from the leaf of vegetation. This is simply what is meant by the proposition that all these organs are transformed or metamorphosed leaves. What would have been leaves, if the development had gone on as a vegetative branch, have in the blossom developed in other forms, adapted to other functions. Linnæus expressed this idea, along with other more speculative conceptions, dimly apprehended, by the phrase Vegetable Metamorphosis. Not long afterwards, this fecund idea of a common type, the leaf, of which the parts of the flower were regarded as modifications, was more clearly and differently developed by a philosophical physiologist, Caspar Frederic Wolff. Thirty years later it was again and wholly independently developed by Gœthe, in a long-neglected but now well-known essay on the Metamorphoses of Plants. Twenty-three years afterwards, similar ideas were again independently propounded by DeCandolle, from a different theoretical point of view."

The author of "Faust" and of "Wilhelm Meister" was equally at home in botanical and anatomical studies, and bids fair to be remembered by his essay already mentioned and by his discovery of the intermaxillary bone. It was amid such studies that he could acquire that calmness which utters itself in that little verse which Carlyle loved to quote:

>Wie das Gestirn
>Ohne Hast
>Aber ohne Rast
>Drehe sich jeder
>Um die eigne Last.

>"Like as a Star
>That maketh not haste
>That taketh not rest,
>Be each one fulfilling
>His god-given Hest."

Unhasting, unresting, come the flowers of early June. On the hilltop and in the valley, by the roadside and in the heart of the woodland, amid the grass in the meadow and in the tangle of the swamps, they are everywhere seen. June is a wild-flower month. They are then in their greatest abundance. They still have the delicacy of spring, not yet hardened into the coarser vigor of the summer and autumn. The sap is, as it were, not yet thickened. The wonder at the opening springtime has not been exhausted. Our eyes are still attracted by the magic of the colors which lie over the face of Nature. The accompaniment of bird life and insect life is not wanting. Earth, air and sky combine to make this a charming season. The rapidity with which new flowers unfold hardly allows us opportunity to record them all. In a week, at least fifty more are ready in a comparatively small area. Their places will soon be taken by others whose buds are even now swelling, to open when their appointed day comes; for they do not come by chance.

| May 29. | 101 | Osmunda Claytoniana, L. | Interrupted Fern. |
| | | Sanicula Marylandica, L. | Sanicle. |
| | | Actæa alba, Bigel. | White Baneberry. |
| | | Smilacina racemosa, Desf. | False Spikenard. |
| | | Acer Pennsylvanicum, L. | Striped Maple. |
| | | Sassafras officinale, Nees | Sassafras. |
| | | Ranunculus recurvatus, Poir. | Hooked Crowfoot. |
| | 108 | Cardamine hirsuta, L. | Small Bitter Cress. |
| | *1 | Cypripedium pubescens, Willd. | Lady's Slipper. |
| | | Equisetum sylvaticum, L. | Horsetail. |
| | | Quercus coccinea, Wang. | Scarlet Oak. |
| | | Larix Americana, Michx. | American Larch. |
| | | Ornithogalum umbellatum, L. | Star-of-Bethlehem. |

* See note, page 120.

## THE FLOWERS OF EARLY JUNE. I. 151

|  |  |
|---|---|
| Nasturtium Armoracia, Fries | Horseradish. |
| Betula lenta, L. | Black Birch. |
| Betula lutea, Michx. f. | Yellow Birch. |
| Betula papyrifera, Marshall | Paper Birch. |
| x Corallorhiza innata, R. Brown | Coral-root. |
| Quercus Prinus, L. | Chestnut-Oak. |
| Cynoglossum Virginicum, L. | Wild Comfrey. |
| Claytonia Caroliniana, Michx. | Spring-Beauty. |
| Fraxinus sambucifolia, Lam. | Black Ash. |
| Fagus ferruginea, Ait. | American Beech. |
| Ribes prostratum, L'Her. | Fetid Currant. |
| Mitella diphylla, L. | Mitrewort. |
| Tiarella cordifolia, L. | False Mitrewort. |
| Oryzopsis asperifolia, Michx. | Mountain Rice. |
| xx Luzula vernalis, DC. | Wood-Rush. |
| Eriophorum vaginatum, L. | Cotton-Grass. |
| Andromeda polifolia, L. | Andromeda. |
| Kalmia glauca, Ait. | Pale Laurel. |
| Menyanthes trifoliata, L. | Buckbean. |
| Lonicera cærulea, L. | Mountain Honeysuckle. |
| Ledum latifolium, Ait. | Labrador Tea. |
| Allium tricoccum, Ait. | Wild Leek. |
| Ribes floridum, L'Her. | Wild Black Currant. |
| Alopecurus pratensis, L. | Meadow Foxtail. |
| xxx Lonicera glauca, Hill | Honeysuckle. |

June 2. 109 Nepeta Glechoma, Benth.     Ground Ivy.
Lonicera ciliata, Muhl.     Fly-Honeysuckle.
Nemopanthes fascicularis, Raf.     Mountain Holly.

|  |  | Chiogenes serpyllifolia, Salisb. | Creeping Snowberry. |
|---|---|---|---|
|  |  | Rubus triflorus, Richardson | Dwarf Raspberry. |
|  |  | Senecio aureus, L. | Squaw-weed. |
|  |  | Trientalis Americana, Pursh | Star-flower. |
|  |  | Sisyrinchium anceps, Cav. | Blue-eyed Grass. |
|  |  | Clintonia borealis, Raf. | Clintonia. |
| June | 3. | Comandra umbellata, Nutt. | Bastard Toad-flax. |
|  |  | Hypoxis erecta, L. | Star-Grass. |
|  | 120 | Carex cephalophora, Muhl. | Sedge. |
|  |  | Carex canescens, L. | " |
|  |  | Pyrus arbutifolia, L. f. | Choke-berry. |
|  |  | Prunus Virginiana, L. | Choke-cherry. |
|  |  | Ranunculus acris, L. | Tall Buttercups. |
|  |  | Eleocharis tenuis, Schultes | Spike Rush. |
|  |  | Carum Carui, L. | Caraway. |
|  |  | Buda rubra, Dumort. | Sand-Spurrey. |
|  |  | Gaylussacia resinosa, Torr. & Gray | Huckleberry. |
|  |  | Osmunda cinnamomea, L. | Cinnamon Fern. |
|  | 130 | Equisetum limosum, L. | Horsetail. |
| " | 5. | Cratægus coccinea, L. | White Thorn. |
|  |  | Rubus Canadensis, L. | Low Blackberry. |
|  |  | Rubus strigosus, Michx. | Red Raspberry. |
|  |  | Aphyllon uniflorum, Gray | One-flowered Cancer-root. |
|  |  | Berberis vulgaris, L. | Common Barberry. |
|  |  | Carex triceps, Michx., var. hirsuta, Bailey · | Sedge. |

|  |  |  |
|---|---|---|
|  | Carex rosea, Schkuhr | Sedge. |
|  | Carex gracillima, Schwein. | " |
|  | Carex scoparia, Schkuhr | " |
| 140 | Carex straminea, Willd. | " |
|  | Carya amara, Nutt. | Bitternut. |
|  | Cypripedium acaule, Ait. | Stemless Lady's Slipper. |
|  | Rhododendron nudiflorum, Torr. | Swamp Pink. |
|  | Potentilla argentea, L. | Silvery Cinquefoil. |
|  | Quercus alba, L. | White Oak. |
|  | Veronica arvensis, L. | Corn Speedwell. |
|  | Brassica nigra, Koch | Black Mustard. |
|  | Castilleia coccinea, Spreng. | Scarlet Painted-cup. |
| June 6 | Carex debilis, Michx., var. Rudgei, Bailey | Sedge. |
|  | 150 Juglans cinerea, L. | Butternut. |

It will be noticed that nearly a third of the above list are trees or shrubs, that the genus Carex contributes eight species, and that the others are well distributed among different genera. Many of them are worthy of more than a passing glance.

In 1602 Bartholomew Gosnold discovered Cape Cod and, sailing in his small bark around the cape, entered Buzzard's Bay. On one of the islands, named Elizabeth, from his queen, he landed with his men and built a storehouse and fort, intending to lay the foundation of the first New England colony. Trading with the natives on the main land he was soon able to complete a cargo, which consisted chiefly of sassafras-root, then greatly esteemed in England as a sovereign panacea. It is worth remembering that the sassafras-tree, common enough all over this state, furnished by far the largest part of the first cargo sent from Massachusetts to England. At that time "it commanded an extravagant price and treatises were written to celebrate its virtues." It is easily identified by the fragrant smell and spicy taste of the bark and the root. The leaves are characterized by a great variety of form, even on the same twig. Some are found, especially small ones, entire, others two-lobed or three-lobed, of a shape so marked that, once seen, they are not easily forgotten.

Among our small trees there is none more worthy of cultivation for its beauty alone than the striped maple. Its dark-green bark, striped with white lines, attracts the eye, as much as its large leaves and long recurved branches. Its clusters of fruit are also larger than those of any other of our maples, and, as it flowers the last of those in our neighborhood, they ripen the latest and add to its fine appearance in July. Unfortunately it has one bad habit. As it grows older, even before its stem is two inches thick, its beautiful bark cracks and loses its beauty, and the stem dies. Cases are recorded, however, where it has attained a much greater size. It might, by cultivation, undergo such modification through new conditions, that its beauty might be preserved to a good old age.

Of the twenty-five species of birch known to science five are found in this county, the gray, the paper, the black, the yellow and the red. They form one of the most interesting groups of the genus. "No trees are more distinguished for their light and feathery foliage, and the graceful sweep of their limbs. From the delicate and slender gray birch, throwing its thin leaves and often pensile spray lightly on the air, to the broadheaded black birch, with its rich, glossy and abundant foliage, weighing its pendulous branches almost to the ground,—no family affords such a variety of aspect." I like to see them in early spring when covered with their long aments, which hang like tassels of purple and gold among the just opening leaves. Then they yield to none of our forest trees in beauty. They are valuable, too, for their timber; and one of them, the paper birch, was of the greatest importance to the Indians of those regions where it grows naturally, in the making of canoes and tents, both models of ingenuity and taste. Longfellow, in "Hiawatha," represents the building of a canoe, and gives us Hiawatha's appeals to the birch, the cedar, the larch and the fir to contribute of their best. And then we have this pleasing picture of the finished work:

"Thus the Birch Canoe was builded
In the valley, by the river,
In the bosom of the forest;
And the forest's life was in it,
All its mystery and its magic,
All the lightness of the birch-tree,
All the toughness of the cedar,
All the larch's supple sinews;
And it floated on the river
Like a yellow leaf in Autumn,
Like a yellow water-lily."

With the birches I associate the honeysuckles (*Lonicera*), as shrubs of spring. I have found in the neighborhood of this city three of the four species accredited to this state; and I find in Bigelow's "Florula Bostoniensis" of 1840 that the fourth is "said to grow also in Worcester." I am afraid, however, that like the Linnæa and some other rare flowers, it is now extinct in its old habitat here. Perhaps it will some day return.

There are others of the trees and shrubs which are now in bloom that I like to see yearly: among them are the black ash, the beech, the andromeda, the pale laurel, the Labrador tea, the mountain holly, the chokecherry and the barberry. Few persons, unless specially interested, would notice the mountain holly when in bloom; but few could overlook the barberry, so handsome both in flower and in fruit. Yet the mountain holly is associated in my mind with that most delicate of our trailing plants, the creeping snowberry, because both were growing so thickly crowded together in the Auburn swamp in which I first found them.

There is a sheltered nook at the foot of a "Mount Ararat," where many interesting flowers bloom, where I saw some of them for the first time. One of these was the Clintonia. Its large leaves and greenish-yellow flowers render it so conspicuous, that it is still a mystery to me how I could have overlooked it so long. But now it is an integral part of every returning spring, as much as is the trailing arbutus, the dandelion and the daisy. I have found the wild comfrey in only one spot, and I visit that spot for the sake of seeing the pale blue corolla, of a tint unlike that of any other flower I know.

The dainty little Hypoxys too often escapes notice in the masses of cinquefoil and buttercup that stud the meadows with their "subtlest jewelry." On the edge of the woodland the delicate little star-flower is now in its prime. It is ordinarily seven-petalled and seven-stamened, and in that respect almost unique. Perhaps amid the dry soil by the railroad side, or in the rich grass of the near meadow, we may see clusters of the one-flowered cancer-root, a leafless, rather odd-looking plant, a member of an order of root-parasites.

The purple lady's slipper is one of those flowers which I remember from early childhood. The floor of the oak grove behind the little schoolhouse was studded with them. It was many years afterwards before I saw either of the yellow lady's slippers, which are now among my favorite flowers.

The mountain rice is one of the grasses which serves as a reminder of a visit to Mount Wachusett in early spring: with it were found the fetid currant and the wood-rush. I thought I had found all the species of cotton-grass that could possibly be found in this neighborhood, but a friend brought me a new one (*Eriophorum vaginatum*, L.) from North Pond.

It is fair to assume that Bryant had seen the scarlet painted-cup on the prairies of Illinois, for otherwise how could he sing:

> "The fresh savannas of the Sangamon
> Here rise in gentle swells, and the long grass
> Is mixed with rustling hazels. Scarlet tufts
> Are glowing in the green, like flakes of fire;
> The wanderers of the prairie know them well,
> And call that brilliant flower the Painted Cup.
> Now, if thou art a poet, tell me not
> That these bright chalices were tinted thus
> To hold the dew for fairies, when they meet
> On moonlight evenings in the hazel bowers,
> And dance till they are thirsty. Call not up
> Amid this fresh and virgin solitude,
> The faded fancies of an elder world:

But leave these scarlet cups to spotted moths
Of June, and glistening flies, and humming birds
To drink from, when on all these boundless lawns
The morning sun looks hot."

I know a little meadow which is sometimes scarlet with the painted-cup, and the sight of it links me in thought with our poet, and helps me to appreciate the picture which he saw. In my mind's eye are blended now two pictures, one of a little New England meadow, and one of a broad Illinois prairie. If I could see the latter in reality I think it would look familiar merely because of the familiar painted-cup.

# THE FLOWERS OF EARLY JUNE. II.

*June* 1, 1853. To Walden. Clover begins to redden the fields generally. The quail is heard at a distance. Buttercups, of various kinds mingled, yellow the meadows, the tall, the bulbous, the repens. The cinquefoil, in its ascending state, keeping pace with the grass, is now abundant in the fields. This is a feature of June.

—THOREAU—*Summer*.

Thirty years after the date in Thoreau's diary the same kindly Nature which did yield all her shows to please and win that pilgrim wise unrolled for me the same picture in the Millbury meadows. This is one of the secrets of the charm which I find in his diary, that he has had my experience before me, and that a page out of it seems like a page out of my own life. It may be, too, that the diary of another is almost always more interesting than one's own. However that may be, I know that I appreciate more fully Thoreau's work as I recognize his keenness of perception and devout love of all wild Nature by his fidelity to truth where my experience allows me to judge, and I take his word upon trust with all readiness therefore, where his experiences lay out of my range.

On June 13, 1852, he writes: "The *Smilax herbacea*, L., carrion flower, a rank green vine, with long peduncled umbels, small greenish or yellowish flowers, and tendrils just opening, at the Miles swamp. It smells exactly like a dead rat in the wall, and apparently attracts flies like carrion. I find small gnats in it. A very remarkable odor. A single minute flower, in an umbel, open, will scent a whole room. Nature imitates all things in flowers. They are at once the most beautiful and the ugliest objects, the most fragrant, and the most offensive to the nostrils."

A glance in my own diary for 1882 shows me that I found it in bloom on June 12th, and I think I can appreciate fully all that he says of it. The paragraph is not unmeaning to me, as it necessarily must be if I had not seen and smelt the flower.

Under the same date the next year I find this passage, which reminds me of my first conscious sight of a male rose-breasted grosbeak: "What was that rare and beautiful bird in the dark woods under the Cliffs, with black above and white spots and bars, a large triangular blood-red spot on breast, and sides of breast and beneath, white? Note, a warble, like the oriole, but softer and sweeter. It was quite tame. Probably a rose-breasted grosbeak. At first I thought it was a chewink, as it sat sideways to me, but then it turned its breast full toward me, and I saw the large triangular, blood-red spot occupying the greater part of it. . . . It is a memorable event to meet with so rare a bird. Birds answer to flowers, both in their abundance and their rareness. The meeting with a rare and beautiful bird like this is like meeting with some rare and beautiful flower, which you may never find again perchance, like the great purple-fringed orchis, at least. How much it enhances the wildness and the richness of the forest!" I was sitting, on just such a day, on a hillside covered with a young growth of rock chestnut-oak and poplar and striped maple and birch and the various denizens of our New England woodland. A little below me was a great boulder whose top and exposed side were gay with numerous, crowded stems of the pale corydalis (*Corydalis glauca*, Pursh). Since then, under the influence of summer droughts and autumn rains, the thin soil which had been protected by the thick leaves of chestnut and maple, has disappeared, and with it the corydalis. Near by, further down the slope, used to stand a cluster of dark hemlocks; in the shade of which I found my first colony of the small yellow lady's slipper. Miles away to the south and to the east the spires of village churches stood out against the faint, warm sky. A low hum falling on the ear told of the railway traffic that was stirring beyond the hills. While I was listening with an attent ear and watching with patient eye for whatever might come, at the same time enjoying to the full the charming prospect, suddenly there appeared on a low shrub about six feet before me the bird which of all I then wished to see, and which from the descriptions often read I had no difficulty in identifying as a male rose-breasted grosbeak. It was ten years before I saw another, on what I have since called a red-letter day for birds, because within the short space of a quarter of an hour I saw several Baltimore orioles, scarlet tanagers, redwing-blackbirds, my rose-breasted grosbeak, and a whole flock of the commoner birds, heard the song of the oven-bird though I did not see the singer, and heard the drumming of the ruffed grouse.

In the rare June days, the rare birds and the rare flowers come, and yet these are not all. It is the common things that make up the most of the world, the common birds and the common flowers. These last now light up many a bit of woodland or grassy meadow or bank of quiet pool or babbling brook with their fleeting beauty, and then, having done their utmost, quietly give place to others which in varying forms repeat the old, old story. These belong equally to us all; the wild flowers are not to be monopolized any more than the air of heaven.

When the rhodora and the pink azalea have gone by, there comes a crowd of white-robed shrubs to adorn the untrimmed roadside and the forest paths, with here and there a faintly-tinted companion, as when the mountain laurel bursts forth in its well-appreciated beauty. The buttercups, the dandelions, the ragworts and others contribute a mass of yellow, the painted-cup furnishes with its scarlet a bright contrast to the green of the grasses, and the purple arethusas, in their modest way, supplement the picture. The whole is full of harmony; and, as the scene shifts or the point of view changes, the harmony is preserved. There is no too little, no too much. There is a place for everything, and everything is in its place. The lordly trees may appropriate large spaces for themselves, but the humble mosses will grow in the moist shade beneath, and the leathery lichens will cover their stems and branches; birds will nest there, and insects will burrow in the bark, so that the single tree becomes a colony of life.

But the fourth division of our list is ready. It contains some interesting plants, which are worthy of more than a passing glance.

| | | | |
|---|---|---|---|
| June 6. | 151 | Trifolium repens, L. | White Clover. |
| | | Trifolium pratense, L. | Red Clover. |
| | | Carex pallescens, L. | Sedge. |
| " | 7. | Lupinus perennis, L. | Wild Lupine. |

## THE FLOWERS OF EARLY JUNE. II.

|  |  |  |
|---|---|---|
| | Cypripedium parviflorum, Salisb. | Smaller Yellow Lady's Slipper. |
| | Pogonia verticillata, Nutt. | Pogonia. |
| | Cornus Canadensis, L. | Bunchberry. |
| | Malva rotundifolia, L. | Common Mallow. |
| | Achillea Millefolium, L. | Common Yarrow. |
| 160 | Carya alba, Nutt. | Shag-bark Hickory. |
| | Carex intumescens, Rudge | Sedge. |
| | Carex folliculata, L. | " |
| | Carex polytrichoides, Muhl. | " |
| | Nasturtium officinale, R. Br. | Water-Cress. |
| | Prunus serotina, Ehrh. | Wild Black Cherry. |
| June 9. | Xanthoxylum Americanum, Mill. | Prickly Ash. |
| | Rubus villosus, Ait. | High Blackberry. |
| | Rubus occidentalis, L. | Black Raspberry. |
| | Iris versicolor, L. | Blue Flag. |
| 170 | Quercus prinoides, Willd. | Little Chestnut Oak. |
| | Quercus ilicifolia, Wang. | Scrub Oak. |
| | Geum rivale, L. | Purple Avens. |
| | Ranunculus multifidus, Pursh | Buttercup. |
| | Cornus florida, L. | Flowering Dogwood. |
| | Cornus alternifolia, L. f. | Cornel. |
| | Poa pratensis, L. | June Grass. |
| | Veratrum viride, Ait. | Indian Poke. |
| " 10. | Chrysanthemum Leucanthemum, L. | Ox-eye Daisy. |
| | Dactylis glomerata, L. | Orchard Grass. |
| 180 | Podophyllum peltatum, L. | Mandrake. |
| | Carex tetanica, Schkuhr | Sedge. |

|  |  |  |
|---|---|---|
| | Carex crinita, Lam. | Sedge. |
| | Phleum pratense, L. | Timothy. |
| | Viburnum Lentago, L. | Sweet Viburnum. |
| | Quercus coccinea, Wang., var. tinctoria, Gray Black Oak. | |
| | Salix nigra, Marsh. | Black Willow. |
| | Osmunda regalis, L. | Flowering Fern. |
| June 12. | Arethusa bulbosa, L. | Arethusa. |
| | Smilax herbacea, L. | Carrion Flower. |
| 190 | Rhus Toxicodendron, L. | Poison Ivy. |
| | Habenaria Hookeri, Torr. | Habenaria. |
| | Botrychium Virginianum, Swartz | Moonwort. |
| " 13. | Solanum Dulcamara, L. | Bittersweet. |
| | Myrica cerifera, L. | Bayberry. |
| | Carex echinata, Murray, var., microstachys, Boeckl. | Sedge. |
| | Carex livida, Wahl. | " |
| | Glyceria acutiflora, Torr. | Manna Grass. |
| | Alopecurus geniculatus, L., var., aristulatus, Torr. | Floating Foxtail Grass. |
| | Corydalis glauca, Pursh | Pale Corydalis. |
| 200 | Leucothoe racemosa, Gray | Leucothoe. |

It will be noticed that fifteen of these are trees or shrubs, that there are eight sedges as in the last list, and five grasses. June may be the month of roses, but it is certainly as much the month of grasses. As they wave in the gentle June breezes they seem to fill every nook of the landscape not otherwise occupied. The *Poa*, the

*Leucothoë racemosa*, Gray.
*The Leucothoë.*

*Thou hast few rivals, rare Leucothoë*
*In grace and loveliness among the flowers.*

*Dactylis* and the *Phleum* are widely cultivated and are, for us, three of the most valuable members of this great family, excepting the cereals. The *Glyceria* and the *Alopecurus*, of differing habit and habitat, are wild forms which hover on the borders of cultivation but are not of it.

The two clovers and the lupine are gregarious plants and sometimes tinge whole fields with their own color. In the meadows the clovers may usurp the place of the grasses and yet they are not considered intruders; rather they are sought and cultivated as two of the most valuable fodder plants. Other plants of the same order, the Leguminosæ, are attracting attention for the same reason. *Desmodiums*, *Lespedezas* and *Medicagos*, especially the alfalfa (*Medicago sativa*, L.) are now much cultivated, and with profit, in various sections of this country.

Here are four species of orchids, belonging to four different genera, and they are among the most interesting of their kind. The lady's slipper and the *Arethusa* will always attract by their color, and yet the greenish flowers of the *Pogonia* and the *Habenaria* exhibit as exquisite adaptations for the securing of cross-fertilization as any of the order. Once a year, at least, I want to lift the lid-like anther of the *Arethusa* and the *Pogonia* to see if the pollen-masses are yet retained in their snug case, and to touch the button-shaped disk of the pollen-mass of the *Habenaria* and so draw out the entire contents of the anther-cell, as a sphinx moth would probably do if I did not anticipate it. Since the publication of Darwin's "On the Various Contrivances by which Orchids are fertilized by Insects," in 1862, there have been many additions to our knowledge of this subject, but his work stands out preëminent, and as fascinating as a fairy tale. As a reasonably active imagination is necessary for the full enjoyment of the latter, so a little previous knowledge of the subject is wonderfully helpful to our proper appreciation of the former. I read the book after I had become acquainted with many of the local orchids and was delighted with it. I can imagine that the result might have been different if I had read the book first.

On looking in some of my old diaries I see that I have found the *Pogonia verticillata* as early as May 21st, and the *Cypripedium parviflorum* as early as May 22d; but the present season is late. Year after year I found them in Heywood's woods. And not these only; for here is a Luna moth just out of its cocoon, its wings not yet dry; and there runs, close to the ground, a bright olive-green bird which is easily identified as the golden-crowned thrush or oven-bird. A short search reveals a little mound of dead leaves, within which is a nest with three brown-dotted cream-colored eggs. What a solitary place the little oven-bird has chosen for its nest!

Let us hope it may escape the prowlers which infest the woodland. John Burroughs in "Wake Robin" was the first writer to describe to me its songs, but I am not yet certain that I have heard but one.

One of the humble plants, which must find its way into any list of New England plants, is an emigrant hither from the Old World. The common yarrow, however, has an illustrious name (*Achillea Millefolium*). Its connection with the swift-footed Achilles of the Trojan war is certainly mythical, but it pleasantly reminds us that Achilles was a pupil of Chiron, the wisest and justest of all the Centaurs, and skilled in the art of healing.* Under a simple microscope its small heads of flowers form objects of great beauty. Several years ago, on a long, hot climb up Mount Washington by the bridle-path from the Crawford House, I was interested to notice the dwarfing of the vegetation as we ascended the mountain-side, and when we reached the long ridge which slopes away up to the topmost peak, and saw all around only dwarf birches and willows, lichens and mosses, we realized we were in the sub-arctic zone of vegetation as truly as if we were two thousand miles further northward. But at the very summit, a vigorous plant of *Achillea* was in bloom, undaunted as its namesake in the midst of dangers that

---

* See Landor's *Imaginary Conversations*, Vol. I., Achilles and Helena.

beset it around. The yarrow now seen by the roadside and that gathered on the summit of Mount Washington on that August day are closely linked together in my thought as an illustration of the capacity some plants have of adapting themselves to a new environment.

Near the railroad crossing on the road from Millbury to Wilkinsonville is the one place where I have found our one representative of the order Rutaceæ, most of which are natives of South America and the temperate climes of other lands. This is the northern prickly ash, not at all connected, except in name, with our other ash-trees. Its branches are covered with strong, sharp prickles, arranged in no definite order, and the leaves are pinnate. The bark is bitter, aromatic and stimulant, sometimes used to alleviate the toothache; from which fact it is sometimes called toothache-tree.

The shagbark or shellbark hickory is easily the chief of all the trees in our list. The genus is an exclusively North American one. The uses to which its wood is put are very numerous. As fuel, it stands at the head of all trees in our climate. It is the heaviest of our native woods and yields, cord for cord, more heat than any other, in any shape in which it may be consumed. Its specific gravity is .8372; its relative approximate fuel value is .8311. It is worthy of cultivation for its nuts alone.

The poison ivy (*Rhus Toxicodendron*, L.) is one of the few plants of this neighborhood poisonous to the touch, although all persons are not affected by it. It is easily identified by the three leaflets forming a compound leaf. Its blossoms show its kinship with the sumachs. The bittersweet (*Solanum Dulcamara*, L.) proves its kinship with the potato as effectively by its blossoms, and by the fact that both are a prey to the same enemy, the Colorado beetle. In the middle of a swamp, far from any cultivated field, the beetle finds the bittersweet, and unterrified by Paris green feeds in quiet and multiplies unseen.

In the low meadows the nodding flowers of the purple avens contrast finely with the brighter colors around. The surface of this little pool is covered with yellow flowers which prove, on closer inspection, to be buttercups, but with what strange leaves under the water! On the upland the ox-eye daisy shines afar. But these are not all; other flowers fill other spaces, each contributing its mite to clothe the earth with beauty. It is a pleasure to me that I have been able to see so many of them in other years, so that now I can welcome them as old friends in the old familiar places.

# THE MID-JUNE FLOWERS.

Happy, truly, is the naturalist. He has no time for melancholy dreams. The earth becomes to him transparent; everywhere he sees significancies, harmonies, laws, chains of cause and effect endlessly interlinked, which draw him out of the narrow sphere of self-interest and self-pleasing, into a pure and wholesome region of solemn joy and wonder.

—KINGSLEY— *Glaucus.*

The mid-June grasses, tall and lush, used to bend over a brook, a wonderful brook it seemed to be then, which came from somewhere beyond the low hills which formed the Ultima Thule of the boys' world, and flowed through the wide meadow to join the great river. There was a current rumor among the boys that the river flowed into the ocean somewhere, but none of them had seen the ocean. A section of the brook and a section of the river formed in summer a large part of their little world, and they did not trouble themselves about the mysteries of the source of the one or the remote course of the other. It was sufficient for them that the meadow and the brook and the river were overflowing with a life like their own. They wandered through the grass in the meadow, and saw it wave in the gentle breeze; they startled the fish in the brook and ran along the bank to see more of their wonderful life. They watched the water plants rise and fall with the current, and delighted in the rippling music of the brook where it slipped hurriedly over the pebbles. Sometimes they crept with stealthy tread to the margin with short poles and pieces of wire to snare the little pickerel which lay so motionless near the surface, but which awoke all too soon to a sense of danger and darted out of sight. They ran, hat in hand, after the butterflies which rose from the flowers amid the grass; they hunted for the nests of birds which fluttered timidly before them. They delighted in seeing frogs dive into the brook and swim under the bank for shelter, and they poked them out of their hiding-places to watch their frantic efforts to escape.

Life was all about them and in them. The sunshine filled their little world with brightness. Light was life to them. Consciously or unconsciously, they reveled in both. The days were long, because they were so full of pleasure, and the seasons and the years seemed almost endless.

There were times when the boys tired of the brook and climbed the railroad embankment that bordered the meadow and separated it from the great river. There they watched men fishing, who caught great strings of yellow perch and dace and pickerel; and they took their first lessons in that noble art. Best of all, they were learning to love with an abiding love the wild things of the field, the wood, the water and the air, a love which afterwards developed into reverence for the wonderful Nature around them. Bird and insect and flower entered into their life, and have since added no small charm to the pleasure of existence, for the interest in these things has not been lost amid the toil and the coil incident to a busy, earnest manhood. Books were scarcer in those days than they are now, but somehow the boys learned to know many things which were handed down from preceding generations of boys by oral tradition. They did not know all the birds by name, but they had names for all the common ones and knew their habits. Their quick eyes could detect a night-hawk sitting along a high limb, and they could dissolve any doubt about it by hitting the limb with a stone if one was near, or by jarring the tree, to make the bird fly. They knew where to look for a pigeon-woodpecker's nest, how to get a chipmunk out of his hole, and the places along the river where you could be sure to get a good string of fish.

Most of the lore learned in this way is not forgotten. It is usually learned by heart and becomes a part of one's nature: It helps us, also, very much to appreciate book-lore, and, perhaps, aids in developing a love for good literature. Its influence is certainly on the right side.

In these mid-June days the woods stand forth in their glory, and are seen at their best when the westering sun gleams athwart them, lighting them up with a blaze of yellow, while the valley below lies enshrouded in the dark shadow. How much the trees contribute to the beauty of a New England landscape can be comprehended best by those who are familiar with treeless regions. It is not difficult to realize how easily in old days, when the world was young and our race had still the imagination of a child, nymph and dryad might people the forest shades. Our stout old Germanic ancestors learned and practised in the dim depths of their primeval forests many of those virtues which make their descendants to this day the foremost nations of the world. There the unconquerable spirit of personal liberty was born. In such a spot a poet might sing:

"Here are old trees, tall oaks and gnarled pines,
  That stream with gray-green mosses; here the ground
  Was never trenched by spade, and flowers spring up
  Unsown, and die ungathered. It is sweet
  To linger here, among the flitting birds,
  And leaping squirrels, wandering brooks, and winds
  That shake the leaves, and scatter as they pass,
  A fragrance from the cedars, thickly set
  With pale-blue berries. In these peaceful shades—

Peaceful, unpruned, immeasurably old —.
My thoughts go up the long, dim path of years,
Back to the earliest days of liberty."

Now come rare flowers, rare in number, rare in gracefulness, the full perfection of whose charm can be realized only in their native haunts, by the now faintly murmuring brook, or within the shadow of the pine or the maple. Like the golden opportunity they present themselves and then pass away; but unlike the golden opportunity they will come again with the recurring seasons, as fresh and fair as a first creation, realizing as nearly as may be the fanciful dream of the "Fountain of Youth."

Our fifth group of flowers has been hurrying on. They seem impatient to take their places in the long summer procession. We can hardly keep pace with them in their forward movement. Some of them are so beautiful, we are loth to let them pass. All of them have an interest for us as finished specimens from Nature's workshop. If we cannot stop to examine each minutely, we can recognize their presence and hope to see them again and, perhaps, learn more of their life history. At least we will welcome them as old friends.

## THE MID-JUNE FLOWERS.

| June 14. | 201 | Nyssa sylvatica, Marsh. | Tupelo. |
| | | Gaylussacia frondosa, Torr. & Gray | Dangle-berry. |
| " 15. | | Viburnum acerifolium, L. | Arrow-wood. |
| | | Medeola Virginiana, L. | Indian Cucumber-root. |
| " 16. | | Callitriche verna, L. | Water Starwort. |
| | | Kalmia latifolia, L. | Mountain Laurel. |
| | | Heracleum lanatum, Michx. | Cow-Parsnip. |
| | | Galium triflorum, Michx. | Sweet-scented Bedstraw. |
| " 17. | | Plantago lanceolata, L. | Ribgrass. |
| | 210 | Helianthemum Canadense, Michx. | Frost-weed. |
| | | Nuphar advena, Ait. f. | Spatter-Dock. |
| | | Viburnum cassinoides, L. | Withe-rod. |
| | | Carex riparia, W. Curtis | Sedge. |
| " 19. | | Rumex crispus, L. | Curled Dock. |
| | | Festuca elatior, L. | Meadow Fescue. |
| | | Lepidium Virginicum, L. | Wild Peppergrass. |
| | | Silene Cucubalus, Wibel | Bladder Campion. |
| " 20. | | Trifolium agrarium, L. | Yellow Clover. |
| | | Kalmia angustifolia, L. | Sheep Laurel. |
| | 220 | Geum Virginianum, L. | Avens. |
| | | Viburnum Opulus, L. | Cranberry-tree. |
| | | Glyceria nervata, Trin. | Fowl Meadow-Grass. |
| " 21. | | Lolium perenne, L. | Common Darnel. |
| | | Carex vulpinoidea, Michx. | Sedge. |
| | | Hieracium venosum, L. | Rattlesnake-weed. |
| | | Nasturtium palustre, DC. | Marsh Cress. |
| | | Eleocharis palustris, R. Br. | Spike-Rush. |
| | | Sarracenia purpurea, L. | Pitcher-Plant. |

|  |  |  |
|---|---|---|
|  | Melampyrum Americanum, Michx. | Cow-Wheat. |
|  | 230 Œnothera pumila, L. | Evening Primrose. |
|  | Brunella vulgaris, L. | Self-heal. |
|  | Danthonia spicata, Beauv. | Wild Oat-Grass. |
|  | Galium trifidum, L. | Small Bedstraw. |
|  | Potentilla Norvegica, L. | Cinque-foil. |
| June 23. | Erigeron annuus, Pers. | Daisy Fleabane. |
|  | Lobelia spicata, Lam. | Lobelia. |
|  | Angelica atropurpurea, L. | Angelica. |
|  | Stellaria longifolia, Muhl. | Long-leaved Stitchwort. |
|  | Lysimachia quadrifolia, L. | Loosestrife. |
|  | 240 Liparis liliifolia, Richard | Twayblade. |
|  | Agropyrum repens, Beauv. | Quitch-Grass. |
|  | Allium Canadense, Kalm | Wild Garlic. |
|  | Briza media, L. | Quaking Grass. |
|  | Polygonum sagittatum, L. | Arrow-leaved Tear-thumb. |
|  | Celastrus scandens, L. | Climbing Bittersweet. |
|  | Diervilla trifida, Moench | Bush Honeysuckle. |
|  | Bromus racemosus, L. | Upright Chess. |
|  | Phalaris arundinacea, L. | Reed Canary-Grass. |
|  | Agrostis alba, L. | White Bent-Grass. |
|  | 250 Acer spicatum, Lam. | Mountain Maple. |

Only ten of this list are trees or shrubs, which is a smaller proportion than usual thus far. There are only three sedges, but there are eight grasses. Three viburnums come quite near together, all being in bloom at the same time, and often in the same locality. The cranberry-tree is here often transplanted to the lawn to add its charm to the exotics, its companions. It is not abundant here, at least I have not found it so, though it is found in great profusion further north; so common is it in Aroostook County in Maine that a writer in "The American Naturalist" says, "It made the landscape gorgeous with its scarlet berries."

Among our more picturesque trees the tupelo holds a prominent place. Its top is generally flattened, and its larger branches are covered by a multitude of short twigs at right angles to the branch. Its foliage is glossy green; and, when once identified, it is a favorite tree amid rude, untamed scenery. It loves best moist situations, such as the margins of ponds, although it is sometimes found standing alone on a bare hillside. When clad in its autumnal colors it is one of the brightest ornaments of that season.

The mountain laurel now holds the first place among the shrubs. The flagging interest in the wild flowers — if it is possible for the interest in them to flag — is sure to be revived by the sight of these gorgeous pink and white clusters which fill the pastures and the open woodlands. The sheep laurel is so overshadowed by its more favored brother that it does not get all the credit it deserves for its efforts to beautify the earth in its own way.

I came near missing the little water starwort when in bloom, because of its inconspicuous flowers, which consist of a single stamen and a single ovary without calyx or corolla, almost hidden in the axil of one of the leaves. I look for it year after year in the same brook in which I first found it, and it never fails to appear. The song of Tennyson's brook it has made its own in that one respect at least.

More and more of the marsh plants are now coming into bloom, one of which deserves and attracts more than a passing interest. This is the pitcher-plant, named by Tournefort, the leading French botanist at the end of the seventeenth century, in honor of Dr. Sarrazin of Quebec, who first sent our species, *Sarracenia purpurea*, L., and a botanical account of it, to Europe. In a note to that charming work, "The Old Régime in Canada," by Francis Parkman, I find this brief account of Dr. Sarrazin, who was one of the few Frenchmen of a certain intellectual eminence then living in Canada. "Sarrazin, a naturalist as well as a physician, has left his name to the botanical genus *Sarracenia*, of which the curious American species, *S. purpurea*, the 'pitcher-plant,' was described by him. His position in the colony was singular and characteristic. He got little or no pay from his patients; and, though at one time the only genuine physician in Canada, he was dependent on the king for support. In 1699, we find him thanking his

Majesty for 300 francs a year, and asking at the same time for more, as he had nothing else to live on. Two years later the governor writes that, as he serves almost everybody without fees, he ought to have another 300 francs. The additional 300 francs was given him; but, finding it insufficient, he wanted to leave the colony. 'He is too useful,' writes the governor again; 'we cannot let him go.' His yearly pittance of 600 francs was at one time reënforced by his salary as member of the Superior Council. He died at Quebec in 1734." His name, however, shall not fade, cannot fade, from human remembrance so long as men study the wild flowers of America, where, I believe, this genus is as yet exclusively found.

The pitcher-shaped leaves, usually half filled with water and drowned insects, with their rounded arching hood at the apex clothed with stiff bristles pointing downward, making the entrance easy for insects but the exit almost impossible, when once seen and examined ever so slightly, are not to be forgotten. In one respect the flower also is peculiar. The short style of the pistil is expanded at the summit into a broad 5-angled, 5-rayed, umbrella-shaped body, about as large as a horizontal section of the whole flower, while the five delicate rays terminate under the angles in as many little hooked stigmas. This forms a cover for the numerous stamens, which are not visible unless the petals be drawn aside. Once, when examining some of these flowers while still growing, I was surprised to find this cavity filled with flies somewhat larger than the common house-fly, all busy as could be in eating the pollen, of which scarcely a grain then remained. I counted fourteen flies in one flower. Nearly every one examined was filled in the same way. There was a shower coming up at the time, but the flies were evidently seeking food, if not shelter. In the leaf their fate would have been very different.

When I want to find it in great abundance I go to the margin of a pool in the depths of a cedar swamp in Auburn. For convenience, when referring to this pool, I call it "The Lake of the Dismal Swamp." There in the sphagnum moss, where at every step the feet sink out of sight and we wonder what may be beneath, the pitcher-plant is at home. If I ever visit the real "Lake of the Dismal Swamp" I shall expect to find there one of the allied species of this interesting genus.

The cow-parsnip and the great angelica will not be overlooked. They overtop most of their associates and challenge attention by their stature, if in no other way. The delicate little rock-rose sheds its petals so quickly that we can hardly see the use of its opening them, while its neighbor in the list, the dog lily, is persistent in retaining its petals as long as possible. The little evening primrose now opens its quiet eye, and is liable to be confounded, at a hasty glance, with the cinquefoil, its neighbor. The lobelia with its small blue flowers would scarcely suggest any possibility of relationship with the bright, tall cardinal flower of a later date, but we shall see that they are near of kin.

June 20th was the date last year for *Liparis*. It is time now to visit the one locality where I have found it. In the meadow through which I pass, the quaking grass (*Briza media*, L.), which rejoices under fifty-four different common names in Britten and Holland's "Dictionary of English Plant-Names," is at its prime. On the margin of the ditch that borders this field are some odorous plants, which prove to be wild garlic; in the edge of the wood, where plant life runs wild, *Lysimachia* and *Celastrus* and *Diervilla* and a host of others are seen. By the side of this now almost obliterated wood-road is our little colony of *Liparis*, perfect in its own beauty, perfect in the charm of its surroundings. There are larger and more gayly-colored flowers to follow, but none which I would rather see than this. I hope to find it here every year, where I first found it, and where I can renew the pleasant surprise of that first discovery.

# THE EARLY SUMMER FLOWERS.

*July 2, 1857.—Calla palustris* with its convolute point, like the cultivated, at the south end of Gowing's swamp. Having found this in one place, I now find it in another. Many an object is not seen, though it falls within the range of our visual ray, because it does not come within the range of our intellectual ray. So in the largest sense we find only the world we look for.
—THOREAU— *Summer.*

As I now read the "Summer" diary of Thoreau and enjoy it thoroughly, I cannot help regretting that it was not printed earlier, so that I might have had it as a guide when my own interest in Nature was awaking. One can certainly see some things by himself, but he can as certainly see many more if led by a sympathetic and interested guide. There are not many avenues of human thought or industry where the finger-posts set up by those who have already traveled the road are not helpful. I find that I like to know what other men think of those things which interest me. And so outdoor books, as they may be called, books which present

Nature as viewed by the naturalist rather than by the biologist, naturally form a part of my library.

So I find it pleasant to wander in imagination by the side of gentle Izaak Walton through the sweet English meadows, listening to the song of lark and nightingale and milkmaid, breathing the delicious odor of the hawthorn hedges, filling my basket with the finny treasures of the stream and filling my mind with sweetness and light, as he discourses on the old-fashioned but choicely-good poetry of Kit Marlow and Sir Walter Raleigh and George Herbert and Edmund Waller and other poets of the time. It does not matter now that the modern angler studies his art in other books or from other masters, the perennial charm of the style of "The Compleat Angler," which has stood the test of nearly 250 years, has gained for its author a literary immortality.

It is now a little more than a hundred years since Gilbert White's "Natural History of Selborne" was first printed, perhaps the most famous book of its kind in any language. The feeling of many readers toward it has been best expressed by John Burroughs in his "Indoor Studies." He says: "I was moved to take down my White's 'Selborne' and examine it again for the source of delight I had had in it, on hearing a distinguished literary man, the late Richard Grant White, say it was a book he could not read with any degree of pleasure; to him its pages were a bare record of uninteresting facts. . . . There is indeed something a little disappointing in White's book when one takes it up for the first time, with his mind full of its great fame. When I myself first looked into it many years ago, I found nothing in it that attracted me, and so passed it by. Much more recently it fell into my hands, when I felt its charm and value at once. As a stimulus and spur to the study of natural history it has no doubt had more influence than any other work of the century. Its merits in this direction alone would perhaps account for its success. But, while it has other merits, and great ones, it has been a fortunate book; it has had little competition; it has had the wind always with it, so to speak. It furnished a staple the demand for which was always steady and the supply small. There was no other book of any merit like it for nearly a hundred years. It does not appeal to a large class of readers, and yet no library is complete without it."

Not very far from the parish of Selborne, where Gilbert White lived and wrote, is the parish of Boldre, where William Gilpin, its vicar, was living at the same time and was writing his "Remarks on Forest Scenery," a work which has passed through many editions. Its scenes are laid in the New Forest in Hampshire, with every part of which Gilpin was familiar, and with the life in it he was in entire sympathy. I prize my own copy of this work the more highly because it is of the rare third edition of 1808, in two volumes, a copy of which Francis George Heath, a writer on kindred topics and editor of Gilpin's book, says he could not find in the British Museum.

In the adjacent county of Wiltshire was the home of Richard Jefferies, the best word-painter of rural life in England that this century has seen. His books hold a unique place in the literature of Nature. For a sample of his skill the story of the trout in "Nature near London" and "The Pageant of Summer" in "The Life of the Fields" will suffice; but every one of the dozen volumes is filled with a charm which perhaps no other man could give to them.

But we do not need to go so far from home to find genuine lovers of Nature skilled in the art of expressing their love. What White did for Selborne and Jefferies for Coate was done as effectually, but entirely in his own way, by Thoreau. As Aias stood preëminent among the Argives by the measure of his head and broad shoulders, so stands Thoreau among men who have loved Nature. He stands alone, not to be compared with others, for he is incomparable. It does not help us to appreciate him, to call him the American Gilbert White or Richard Jefferies, any more than it helps us to appreciate Jefferies to call him the English Thoreau. Concord and Walden and the Maine Woods have had in him their prose-poet, and the round of seasons their faithful chronicler.

But while I have been tarrying in the flowery paths of one of my favorite fields of literature, the early summer flowers have been blooming about my feet, and present a brave array.

| | | | |
|---|---|---|---|
| June 24. | 251 | Carya porcina, Nutt. | Pignut Hickory. |
| | | Apocynum androsæmifolium, L. | Dogbane. |
| | | Polygonum cilinode, Michx. | Bindweed. |
| " 26. | | Panicum latifolium, L. | Panic-Grass. |
| | | Scirpus atrovirens, Muhl. | Club-Rush. |
| | | Vitis Labrusca, L. | Northern Fox-Grape. |
| | | Thalictrum polygamum, Muhl. | Tall Meadow-Rue. |
| | | Pinus Strobus, L. | White Pine. |
| | | Arrhenatherum avenaceum, Beauv. | Oat-Grass. |
| | 260 | Raphanus Raphanistrum, L. | Wild Radish. |
| | | Linnæa borealis, Gronov. | Twin-flower. |
| " 28. | | Calla palustris, L. | Wild Calla. |
| | | Myosotis laxa, Lehm. | Forget-me-not. |
| | | Viburnum dentatum, L. | Arrow-wood. |
| | | Habenaria fimbriata, R. Br. | Fringed-Orchis. |
| | | Pogonia ophioglossoides, Nutt. | Pogonia. |
| | | Erigeron strigosus, Muhl. | Daisy Fleabane. |
| | | Oxalis corniculata, L., var. stricta, Sav. | Yellow Wood-Sorrel. |
| " 29. | | Hypericum perforatum, L. | St. John's-wort. |
| | 270 | Hieracium Gronovii, L. | Hairy Hawkweed. |

## THE EARLY SUMMER FLOWERS. 189

|  | Andromeda ligustrina, Muhl. | Andromeda. |
|---|---|---|
|  | Cornus paniculata, L'Her. | Panicled Cornel. |
|  | Rudbeckia hirta, L. | Cone-flower. |
| 274 | Ilex verticillata, Gray | Black Alder. |
| XXXI | Eriophorum polystachyon, L. | Cotton-Grass. |
|  | Eriophorum polystachyon, L., var. latifolium, Gray | " " |
|  | Sagina procumbens, L. | Pearlwort. |
|  | Rhamnus cathartica, L. | Common Buckthorn. |
|  | Smilax rotundifolia, L. | Common Greenbrier. |
|  | Cornus circinata, L'Her. | Round-leaved Cornel. |
|  | Stellaria borealis, Bigel. | Northern Starwort. |
|  | Vaccinium Oxycoccus, L. | Small Cranberry. |
|  | Asclepias quadrifolia, L. | Milkweed. |
| XL | Cynoglossum officinale, L. | Hound's Tongue. |
|  | Convolvulus spithamæus, L. | Bracted Bindweed. |
|  | Smilacina trifolia, Desf. | False Solomon's Seal. |
|  | Medicago sativa, L. | Alfalfa. |
|  | Penstemon pubescens, Solander | Beard-tongue. |
|  | Potentilla tridentata, Ait. | Three-toothed Cinquefoil. |
|  | Linaria Canadensis, Dumont | Toad-Flax. |
|  | Moneses grandiflora, Salisb. | One-flowered Pyrola. |
|  | Sedum acre, L. | Mossy Stone-crop. |
|  | Calamintha Clinopodium, Benth. | Basil. |
| L | Pyrus Americana, DC. | Mountain Ash. |

July 1. 275 Rosa lucida, Ehrh.     Wild Rose.
Spiræa salicifolia, L.     Meadow Sweet.
Rhus venenata, DC.     Poison Dogwood.
Rubus hispidus, L.     Swamp Blackberry.
Panicum dichotomum, L.     Panic-Grass.
280 Panicum depauperatum, Muhl.     " "
Apocynum cannabinum, L.     Indian Hemp.
" 4. Sambucus Canadensis, L.     Common Elder.
Cornus sericea, L.     Silky Cornel.
Eriophorum Virginicum, L.     Cotton-Grass.
Rosa rubiginosa, L.     Sweetbrier.
Calopogon pulchellus, R. Br.     Calopogon.
Pastinaca sativa, L.     Parsnip.
Asclepias Cornuti, Decaisne     Common Milk-weed.
Asclepias phytolaccoides, Pursh     Poke Milk-weed.
290 Asclepias purpurascens, L.     Purple Milkweed.
Erigeron strigosus, Muhl. var. discoideus, Robbins     Daisy Fleabane.
Anthemis Cotula, DC.     May-weed.
Leonurus Cardiaca, L.     Motherwort.
Verbascum Thapsus, L.     Mullein.
Œnothera biennis, L.     Evening Primrose.
Leontodon autumnalis, L.     Fall Dandelion.
Typha latifolia, L.     Cat-tail.
Epilobium angustifolium, L.     Willow-herb.
Conium maculatum, L.     Poison Hemlock.
300 Calamagrostis Canadensis, Beauv.     Blue-Joint Grass.

In this list of seventy, more than one fifth are trees and shrubs; one of them, the white pine, or as it is called in England, the Weymouth pine, is the tallest and most stately tree of our forests. It rises in a single straight column, tapering gradually, to a height of one hundred feet and more. It has a wide geographical range in North America, from the Saskatchewan river in $54°$ north latitude to the slopes of the Blue Ridge in Georgia, from Nova Scotia to the Pacific. It is found everywhere in New England, in every variety of soil. It is the most useful and indispensable of our trees, affording a timber of very great value for many purposes. Yet, as Wilson Flagg says, it has no legendary history, because it is an American tree. Ralph Waldo Emerson has sung its praises in his "Woodnotes." It is associated with no classical images, like the oak, nor with sacred literature, like the cedar of Lebanon. It is easily distinguished by its leaves being in fives, and by its long cones composed of loosely arranged scales; it should be familiar to every resident of New England.

In *Viburnum dentatum* we have the last of the viburnums, and in *Cornus sericea*, the last of the cornels, groups which have been more or less conspicuous during the past five or six weeks. Both the species of *Apocynum*,—there are only two species of this genus in North America north of Mexico,—are now in bloom.

The fruit, consisting of two long and slender follicles, filled with seeds tipped with long silky hairs, shows their kinship to the milkweeds, which are also well represented in the list.

The sweetbrier is one of the most welcome emigrants from the Old World. There is an odor of very pleasant association about it, apart from its own perfume and fleeting beauty.

The bright petals of the wild rose flash out upon us, to our glad surprise, from many a thicket, and lead us to utter the prayer, "Floreat regina florum." About thirty species of wild roses are now recognized as native of the northern hemisphere, reaching from the Arctic circle as far south as Mexico, Abyssinia and India, while the varieties of cultivated roses are almost countless and are yearly increasing. Our wild rose, though a humble member of the genus, may well be proud of its family connection. For more than two thousand years the rose has been celebrated by the poets; it has been surrounded by the most pleasant legends; it has found a place among the traditions as among the customs of many peoples. Its family was already an old one when Romulus and Remus were drifting down the Tiber to the site of Rome; it is older than the pyramids.

While the genus *Rosa* fills so large a space that its lore is almost a literature, the order *Rosaceæ* fills even a wider space, embracing about a thousand species, and including many valuable cultivated plants. Here belong the delicious fruits of the apple, pear, peach, plum, cherry, quince, apricot, strawberry, blackberry, raspberry, in their numberless varieties. What a space the word *apple*,— for we can hardly believe that it is always the fruit known to us by that name — holds in ancient literature! One illustration will suffice. In far-off Greek days, at the nuptial feast of Peleus and Thetis, to which all the gods and goddesses had been invited except Discord, this divinity, out of revenge at her exclusion, threw a golden apple upon the board, with the inscription, "For the most fair." Out of this incident came a train of circumstances which led to the Trojan war and the Iliad and the Odyssey and the Æneid and, in fact, a whole cycle of ancient and modern poetry. Wordsworth's "Laodamia," Tennyson's "Œnone," Morris's "The Death of Paris" show the influence of the old tale.

Beauty and fragrance and worth are preëminent characteristics of this order, though not shared in by all its members equally. Many of the valuable species are at home in our county, either wild or cultivated. There is no fairer sight to be seen in mid-May than an apple-orchard in perfection of bloom; if there is, it must be the same orchard in perfection of fruit.

Beside the two roses in the list, there are the meadow-sweet, a little shrub well worthy of cultivation, and the mountain-ash, one of our most ornamental trees, and a small running blackberry, representatives of our thirty-six or thirty-seven species.

By roadsides the whiteweed now begins to be abundant, and on the edge of swamps the cat-tail flag is blooming. In the low meadows, especially on the edges of ditches, the poison hemlock opens its small white flowers, under the shade, it may be, of the elder or the poison dogwood. There are three bright-colored orchids in the list, the little *Pogonia,* the *Calopogon* and the larger, more majestic *Habenaria,* the latter being one of our handsomest species, and well upholding the reputation of the order for novelty and beauty. The grasses and sedges play a larger part in filling the landscape, especially those species not valued for forage. Whereever moisture is, there these are abundant. Without them the early summer landscape would lose an indefinable and, perhaps, unrecognized charm.

# THE EARLY JULY FLOWERS.

> A pleasing land of drowsyhead it was,
> Of dreams that wave before the half-shut eye;
> And of gay castles in the clouds that pass,
> Forever flushing round a summer sky:
> There eke the soft delights that witchingly
> Instil a wanton sweetness through the breast;
> And the calm pleasures always hover'd nigh;
> But whate'er smack'd of noyance or unrest,
> Was far, far off expell'd from this delicious nest.
> —THOMSON — *The Castle of Indolence.*

The Castle of Indolence lies in fairy Summer Land. The poet's eye, in a fine frenzy rolling, sees it set amid sleep-soothing groves and quiet lawns and flowery beds, and his keen ear listens to the prattle of the purling rills, to the lowing of the herds along the vale and to the flocks loud-bleating from the distant hills: and there rises to our minds as fair a picture of the poet's dream as words can paint. Such pictures belong to summer only. Winter is too serious for such trifling. The stern realities of life are then too apparent, are not to be concealed.

There seems to be something in the nature of summer which incites the mind to linger over such pictures. The bright warm sunlight is in sympathy with the *dolce far niente* spirit in man. Everything about him is growing for him, and why should he not rest and enjoy the fair vision? A stern necessity admonishes him that the hard fates have ordained otherwise. Propt on beds of amaranth and moly in the land of the Lotos, he wearies of the sea, wearies of the oar, merely dreams of fatherland, of wife and child, ceases to think of returning home, until some sage Ulysses leads him back weeping to the hollow ships, and bids to make speed away from the enchanted shore.

But if all the long summer days cannot be given to rest, a part of them can be and ought to be, so that we may see and appreciate the beauty that lies at our feet or before our doors. I have seen from Bethlehem, New Hampshire, the sun sink behind the distant hills, painting the western sky in gorgeous colors and flooding the valley of the Ammonoosuc with purple light; and I have seen from Worcester the sun sink behind Tetaessit Hill in a sky of ineffable beauty, while all the hillside and the valley between were wrapped in richest purple. I have looked down from Mount Washington upon half of New Hampshire spread below, and from Mount Wachusett upon half of Worcester County; the former surpasses in rugged grandeur, but the latter in

## Peat Meadow.

*"But Nature whistled with all her winds,
Did as she pleased and went her way."*
—R. W. EMERSON.

simple beauty: each is altogether lovely in its kind. Lake Winnipesaukee or Lake Memphremagog, girt round with mountains, is a thing of beauty and a joy forever to the lover of wild Nature, but the quiet nooks and wood-clad shores of Lake Quinsigamond have a charm also.

Our love for home sights and sounds should be so great that familiarity cannot breed contempt. Is not the bluebird's matin song, or the bobolink's, as attractive as the skylark's or the nightingale's, if we care to weave about it the romance which early association naturally arouses? It is pleasant, indeed, to sojourn for a time amid foreign scenes, "among new men, strange faces, other minds," but that is not where we would wish to abide. The effect should be to make the home scenes all the dearer. After a day in the Royal Gardens at Kew where the treasures of the Palm House and the North Museum and the Winter Garden are a wonder to see, I find I can still take unalloyed pleasure in roaming along the familiar paths where our simple wild flowers bloom. The latter are all the more interesting from their recognized kinship with those over the sea.

In these early July days, whether we wander in the bright sunlight by the brink of weedy lake, or follow the dainty little brook in its meanderings through the meadows, or seek the refreshing shade of the wood, we shall find new troops of flower friends, some arrayed in dyed garments, glorious in their apparel, others clad in modest robes, as if shrinking to attract our attention. All along our path are the footprints of those which have gone before. Fruit, blossom and bud are all around. There is movement in them all, each striving in its own way to accomplish its own destiny. Even in their decay they will furnish the material out of which new life will spring in other seasons to make the earth pleasant for other men who will then see it. But our seventh list is waiting.

| July | 4. | 301 | Liparis Lœselii, Richard | Twayblade. |
|---|---|---|---|---|
| | | | Woodsia Ilvensis, R. Br. | Woodsia. |
| | | | Cichorium Intybus, L. | Chicory. |
| " | 6. | | Rhus typhina, L. | Staghorn Sumach. |
| | | | Vaccinium macrocarpon, Ait. | Cranberry. |
| | | | Pyrola elliptica, Nutt. | Shin-leaf. |
| | | | Circæa alpina, L. | Enchanter's Nightshade. |
| | | | Silene noctiflora, L. | Night-flowering Campion. |
| | | | Aspidium spinulosum, Swartz, var. intermedium, D. C. Eaton | Wood Fern. |
| | | 310 | Aspidium marginale, Swartz | " " |
| " | 7. | | Aspidium acrostichoides, Swartz | Christmas Fern. |
| | | | Krigia Virginica, Willd. | Dwarf Dandelion. |
| | | | Galium circæzans, Michx. | Wild Liquorice. |
| | | | Pyrola secunda, L. | Wintergreen. |
| | | | Pyrola chlorantha, Swartz | " |

## THE EARLY JULY FLOWERS. 199

|  |  |  |
|---|---|---|
|  | Pyrola rotundifolia, L. | Wintergreen. |
|  | Mitchella repens, L. | Partridge-berry. |
|  | Hydrocotyle Americana, L. | Water Pennywort. |
|  | Trifolium arvense, L. | Rabbit-foot Clover. |
| 320 | Veronica scutellata, L. | Marsh Speedwell. |
|  | Picea nigra, Link | Black Spruce. |
|  | Linaria vulgaris, Mill. | Butter and Eggs. |
|  | Glyceria Canadensis, Trin. | Rattlesnake-Grass. |
| July 8. | Portulaca oleracea, L. | Purslane. |
|  | Plantago major, L. | Common Plantain. |
|  | Holcus lanatus, L. | Velvet Grass. |
|  | Juncus tenuis, Willd. | Rush. |
| " 10. | Carex lupulina, Muhl. | Sedge. |
|  | Carex Pseudo-Cyperus, L., var. Americana, Hochst. | " |
| 330 | Carex bullata, Schkuhr | " |
|  | Scirpus lacustris, L. | Great Bulrush. |
|  | Lysimachia stricta, Ait. | Loosestrife. |
|  | Rhododendron viscosum, Torr. | White Azalea. |
|  | Tephrosia Virginiana, Pers. | Goat's Rue. |
|  | Lilium Philadelphicum, L. | Orange-red Lily. |
|  | Asclepias incarnata, L., var. pulchra, Pers. Swamp Milkweed. | |
|  | Daucus Carota, L. | Wild Carrot. |
|  | Cnicus arvensis, Hoffm. | Canada Thistle. |
|  | Ceanothus Americanus, L. | New Jersey Tea. |
| 340 | Scutellaria galericulata, L. | Skullcap. |
|  | Juncus effusus, L. | Common Rush. |
|  | Juncus acuminatus, Michx. | Rush. |

|  |  |  |
|---|---|---|
| | Campanula aparinoides, Pursh | Marsh Bellflower. |
| | Anemone Virginiana, L. | Anemone. |
| | Sericocarpus conyzoides, Nees | White-topped Aster. |
| | Carex monile, Tuckerm. | Sedge. |
| July 11. | Steironema ciliatum, Raf. | Loosestrife. |
| | Lilium Canadense, L. | Wild Yellow Lily. |
| | Habenaria virescens, Spreng. | Habenaria. |
| 350 | Brachyelytrum aristatum, Beauv. | Beard-Grass. |

Flowering plants have, hitherto, claimed most of our attention, but we should not forget that the flowerless plants fill a great space in the economy of Nature and repay study equally well. Standing among the highest of the Cryptogams, ferns everywhere attract us by their number, variety and beauty. The literature devoted to ferns is very extensive, and some of it very expensive. Perhaps the best single book is Hooker and Baker's "Synopsis Filicum," which describes seventy-five genera and more than 2,200 species. Since this work was first published in 1868 several hundred new species have been discovered and described. My copy of it I value the more highly because it belonged to Dr. Hance, British consul at Amoy, who has enriched it with manuscript notes, being himself an enthusiastic botanist. D. C. Eaton's "Ferns of North America," with its numerous finely colored plates is the best work on our local ferns.

The ferns of the temperate regions vary in height from two or three inches to several feet, but some of the tropical species merit their name of tree-ferns by their height of twenty to twenty-five feet, vying with some of the palms in size and beauty. The ferns are among the first families of the land. They came early, and came to stay. Their fossils, "footprints on the sands of time," are found in rocks from the Devonian Age downward. Coal shales abound with them, sometimes many species being represented on one slab, as perfect as on the day when they were imprisoned. If a living fern could speak, it could hardly tell a plainer tale than these mute relics.

With the naked eye, only the outer beauty of ferns may be seen. The round or elongated spots, brown or black, on the under surface of the frond often pass unobserved. These spots are called *sori*, and are made up of little cup-shaped bodies called *thecæ*. The thecæ are filled with *spores*, from which new plants are developed. All these parts are so minute, except the sori, as to be invisible without the microscope, or at least not to show their true form. But under the compound microscope a whole new world of wonders is laid open. I shall not soon forget the first time that I placed a piece of mature frond of *Aspidium marginale* on the stage of a compound microscope with a two-thirds inch objective. The crowded elastic stems of the thecæ were springing back from their bent position, throwing a shower of spores across the field of vision. I saw the sower sowing the seed with very liberal hand, as Nature always does. About thirty-three species have been thus far reported from this county, and there are, probably, four or five yet to be discovered here. Two-thirds, if not more, of the list may be found in any town.

The trees and shrubs now in flower are only four in number, few as compared with the earlier lists. There are also several evergreen herbs. All the species of *Pyrola* that can be found in this neighborhood might well be found in one day, although I happened to find them on successive days. They form an interesting group, with a strong family resemblance. Species of this genus are found across the continent to the Cascade Mountains. *Pyrola picta*, Smith, from the valley of the Willamette in Oregon, easily proves its relationship with *Pyrola rotundifolia*, L., from the valley of the Blackstone in Massachusetts. In all these species the simple racemes of nodding white or greenish-white flowers cannot fail to arrest attention, and suggest some resemblance, perhaps, to the lily of the valley in the garden.

By the country roadsides the bright blue flowers of the chicory are now opening, a weed fully as handsome as many garden flowers. I have found the little

*Circæa alpina* in two places: one, the bottom of the gorge of "Purgatory" in Sutton, where it is almost buried in the wealth of shield-ferns which abound in those cool moist shades; the other, the "Trossachs" in Scotland, in a similar situation. Having found it in one of these places, I was not surprised to find it in the other. The partridge-berry is a delicate creeper, and when, as it often does, it covers the ground in a mat thickly dotted with the twin flowers, its efforts to contribute to the beauty of the woodland are not to be despised. The water pennywort is another trailing plant, that haunts the sides of the brook or shaded moist grounds. Its bright green leaves are its principal attraction, for the flowers are so small as to escape notice except on careful examination, which reveals the fact that here is a member of the great order Umbelliferæ.

We hardly recognize in this dusty-looking thing by the wayside any relation of the bright clean-looking red clover; but this is the rabbit-foot clover, the outcast of its family. This little pond, shrinking from the gaze of the hot July sun, still furnishes ample nourishment to the many species and numberless individuals which line its borders or creep, now boldly, now timidly, into the water. There we now find blooming the drooping panicles of the *Glyceria*. The large spikes of *Carex lupulina* are visible afar. We are glad to see on the other side of the ditch, the handsome, thickly-set spikes of *Carex Pseudo-Cyperus*, var. *Americana*. What a pity that it has not a simple common name in general use, by which it could be readily identified! *Scirpus lacustris*, with its tall, round, light-green stems, stands with its feet bathed in water, drawing strength from the rich soil, along with several species of *Juncus;* and the swamp-milkweed is a companion for them all.

This clove-like odor comes from that bush of white azalea, the last of its genus to bloom. These fleecy clusters of small white flowers point out the New Jersey tea, which otherwise we might overlook.

But the most showy flowers of this list are the lilies, the wild orange-red, and the yellow. They suggest not delicacy so much as strength, not humility so much as pride. They prefer different situations, the former being found in dry woodlands, the latter, in low meadows. They are the last representatives of the large order Liliaceæ, from which we gladly welcomed in spring the trilliums and the bellworts. They are worthy and fitting to close the long procession. Let us gather a handful of them while we may.

# THE MID-JULY FLOWERS.

> The system of binomial nomenclature as perfected by Linnæus has wonderfully helped the mind of man to domesticate the wild infinity of Nature .... so that a plant-name cannot wander out of the ranks any more than a runaway soldier could elude observation in the ancient Empire of the Cæsars.
> — EARLE — *English Plant Names.*

Natural History studies appeal to the reason, to the judgment, and to the spiritual nature of man. Their pursuit produces in the mind a habit of order and arrangement, and at the same time improves the bodily senses by the accuracy of the observation necessary to discriminate the various objects which pass in review. Their value will depend somewhat upon the method and the fulness with which they are pursued. Suppose children were taught at school nothing but the names of the twenty-six letters of the alphabet and of the ten figures of the Arabic notation, without any insight into the wonderful results to be attained by their manifold combinations and relations, we should undoubtedly say that such study was eminently unpractical, and that the time would be better spent in some more useful occupation. But every child must learn these things in the early part of his school career, and then he gets some insight into the realm of knowledge that lies beyond this open door. How far he will enter in, and how widely he will roam over the Elysian Fields of learning afterwards, depends largely on his own character and opportunities.

It is so with Natural History studies. To acquire the names of plants or animals or minerals, without any idea of their mutual relations, may appear trifling; but to trace the connections and the relations of the various parts of any one of the three great kingdoms of Nature is a pursuit adapted to the highest intellect. What Cicero said of the study of philosophy may be said with equal truth of these studies: "They nourish youth, they delight old age."

The history of botany, like that of any other branch of knowledge, is a history of progress from small beginnings. It is a history of theories adopted as satisfactory for the time being and then discarded, of "obstinate questionings of sense and outward things," of an old order changing and yielding place to new.

The progress of geographical knowledge, the utility of which every one will admit, may be taken as a type of the change which the natural sciences have under-

## THE MID-JULY FLOWERS.       207

gone. A map of the world as known to Homer gives us a comparatively small circle drawn about Greece as a center, including southern Europe, western Asia and northern Africa, the whole surrounded by the Ocean Stream; a geography of myth and fairy tale, of sea gods and nymphs and enchantresses neither mortal nor divine; a geography of strange lands like that Syrian isle "where disease is not, nor hunger nor thirst, and where, when men grow old, Apollo comes with Artemis and slays them with his silver bow."

Little by little the limits of the circle are pushed further back. Herodotus has a larger knowledge than Homer; Ptolemy, than Herodotus; Alexander and Cæsar push their conquests to the verge of the world. In time comes the invention of the mariner's compass, and a bold Columbus pushes westward across the ocean to reach the far East with its boundless wealth, and a new world is laid open to the astonished gaze of mankind, thenceforth stimulated to greater zeal in discovery and exploration and map-making. The coming explorer will soon be obliged to weep for new lands to explore. The names of many who have taken an active part in these discoveries and explorations have become a part of the map, to suggest their history to all students in the after time.

So in a vague sort of way, botany must have been studied from time immemorial. But the beginnings of the science were humble enough, and its progress was slow. A few names stand out prominently in its history, white on a dark background, until the Renaissance gave an impulse to the development of all branches of learning. The first botanists were herbalists; whatever was scientific in the art of medicine was centred in the study of herbs; the remedies of the ancient physicians were of vegetable origin. The mineral and chemical remedies are of a late date; drugs were at first dried herbs. While the beginnings of this science are interesting to the student, the names of Theophrastus and Dioscorides, Pliny, Galen and Apuleius cannot be forgotten.

But the Columbus of botany was Linnæus. He supplied the first necessary condition of all durable progress, by establishing a sound nomenclature; he thoroughly revised the principles of classification and established genera and species upon a more scientific basis. The invention of the microscope has led to the development of certain parts of the subject much as the mariner's compass contributed to the extension of geographical knowledge. Enthusiastic and devoted students have not ceased in their efforts to widen our knowledge of the plant-life of the broad earth, and thus to contribute to the happiness of mankind.

Perhaps no botanists deserve more honorable notice in connection with the flora of our own country than the two Frenchmen, André Michaux and his son,

## THE MID-JULY FLOWERS.

whose names, abbreviated to Mx. or Michx. are appended as the discoverers or describers of many of the plants in my lists. The elder Michaux came to this country in 1785 and traveled here extensively during nine years, collecting indefatigably. In 1803, a year after his death, his "Flora of North America," the first of its kind, was published. His son, who had also resided and traveled here for several years, in addition to other works, published at Paris in 1819, in three volumes, the North American Sylva, "A Description of the Forest Trees of the United States, Canada and Nova Scotia." In 1859, two volumes by Thomas Nuttall were added as a supplement to this work unique in its kind, but now about to be supplanted by the great work of Prof. C. S. Sargent in course of publication.

The two Michaux not only devoted a large part of their lives to the exploration of our forests but showed their abiding love for this study by leaving two legacies for the purpose of encouraging the study of sylviculture in the United States. It is pleasant in this connection to read this brief extract from the will of the younger Michaux, dated September 4, 1855: "Wishing to recognize the services and good reception and the cordial hospitality which my father and myself together and separately have received during our long, and often perilous, travels in all the extent of the United States, as a mark of my lively gratitude, and also to contribute in that country to the extension and progress of agriculture, and more especially of sylviculture in the United States, I give and bequeath to the American Philosophical Society of Philadelphia, of which I have the honor to be a member, the sum of twelve thousand dollars. I give and bequeath to the Society of Agriculture and Arts, in the State of Massachusetts, of which I have the honor to be a member, the sum of eight thousand dollars." The bequest to the Massachusetts Society for the Promotion of Agriculture is applied to aid the botanical garden at Harvard and the Arnold Arboretum, and to the publication of pamphlets on forest culture.

Leaving out the cryptogams of lower rank than the ferns, we find that Michaux's Flora contains 1,530 species, about half the number now described in Gray's "Manual of the Botany of the Northern United States." Eleven years after the publication of Michaux's work, the second Flora of North America, by Frederick Pursh, appeared. This was not confined, like the former, to the author's own collections, but aimed at completeness, and described 3,076 species.

The third Flora, started by Torrey and Gray more than fifty years ago, is still unfinished. The activity of botanical exploration has been so great during this interval, so many new genera and species have been discovered, that we can easily form some just idea of the

## THE MID-JULY FLOWERS. 211

necessary labor involved in the accurate discrimination and proper coördination of the 10,000 or 12,000 species to which the number has been raised, a labor which, in part, accounts for the slowness of the work. The names of Michaux, Pursh and Gray are found in the following list.

| | | | |
|---|---|---|---|
| July 11. | 351 | Rhus glabra, L. | Smooth Sumach. |
| | | Urtica gracilis, Ait. | Nettle. |
| " 13. | | Ophioglossum vulgatum, L. | Adder's-Tongue. |
| | | Nymphæa odorata, Ait. | Water-Lily. |
| | | Gaultheria procumbens, L. | Checkerberry. |
| | | Castanea sativa, Mill., var. Americana, Michx. | Chestnut. |
| | | Microstylis ophioglossoides, Nutt. | Adder's-Mouth. |
| | | Aletris farinosa, L. | Colic-root. |
| | | Aralia hispida, Vent. | Wild Elder. |
| | 360 | Carex trisperma, Dewey | Sedge. |
| | | Polygala polygama, Walt. | Polygala. |
| " 14. | | Asclepias obtusifolia, Michx. | Milkweed. |
| | | Melilotus alba, Lam. | White Melilot. |
| | | Polygala sanguinea, L. | Polygala. |
| | | Nepeta Cataria, L. | Catnip. |
| | | Chenopodium album, L. | Pigweed. |
| | | Lactuca Canadensis, L. | Wild Lettuce. |
| " 17. | | Asclepias tuberosa, L. | Butterfly-weed. |
| | | Pontederia cordata, L. | Pickerel-weed. |
| | 370 | Eriophorum cyperinum, L. | Cotton Grass. |

| July 18. | Drosera rotundifolia, L. | Sundew. |
| | Verbena urticæfolia, L. | White Vervain. |
| " 19. | Verbena hastata, L. | Blue Vervain. |
| | Habenaria lacera, R. Br. | Fringed-Orchis. |
| | Impatiens fulva, Nutt. | Touch-me-not. |
| | Ampelopsis quinquefolia, Michx. | Virginian Creeper. |
| | Galium asprellum, Michx. | Rough Bedstraw. |
| | Mimulus ringens, L. | Monkey-flower. |
| | Eriophorum gracile, Koch | Cotton Grass. |
| 380 | Setaria glauca, Beauv. | Foxtail. |
| | Hypericum ellipticum, Hook. | St. John's-wort. |
| | Utricularia vulgaris, L. | Bladderwort. |
| " 24. | Agrimonia Eupatoria, L. | Agrimony. |
| | Baptisia tinctoria, R. Br. | Wild Indigo. |
| | Saponaria officinalis, L. | Soapwort. |
| | Solidago arguta, Ait. | Golden-rod. |
| | Spiræa tomentosa, L. | Hardhack. |
| | Cyperus filiculmis, Vahl | Galingale. |
| | Bromus ciliatus, L. | Brome-Grass. |
| 390 | Hypericum maculatum, Walt. | St. John's-wort. |
| | Phryma Leptostachya, L. | Lopseed. |
| | Inula Helenium, L. | Elecampane. |
| | Cnicus lanceolatus, Hoffm. | Thistle. |
| | Clematis Virginiana, L. | Virgin's Bower. |
| | Mollugo verticillata, L. | Carpet-weed. |
| | Chimaphila umbellata, Nutt. | Pipsissewa. |
| | Chimaphila maculata, Pursh | Spotted Wintergreen. |

Lycopus sessilifolius, Gray   Water Horehound.
Circæa Lutetiana, L.   Enchanter's Nightshade.
400 Aralia quinquefolia, Decsne. & Planch. Ginseng.

At the head of the list is one of our common shrubs, the smooth sumach. When massed together in good-sized clumps, it plays no insignificant part in the beauty of wayside and upland pasture. Its compound leaves, often a foot or more long, with from thirteen to nineteen leaflets, on a large smooth stalk, serve to identify it easily from the other sumachs. In the low lands, where it can find the moisture that it needs, the hardhack is abundant. The erect position of its stem, crowned by a tapering spire of purple flowers, is said to have gained for it from the early colonists the name of "steeple-bush." Though it begins to flower at the top of its compound panicle, where the flowers are faded before those on the lower branches begin to expand, this little shrub possesses considerable beauty. The *Ampelopsis* is often planted about houses on account of the beauty of its foliage. It need not be mistaken for the poison ivy, if we remember the ivy's three leaflets and the five of the *Ampelopsis*. The clematis, now in bloom, will be more conspicuous by and by when in feathery fruit.

In the grass which covered an old and now disused wood-road in Auburn I once found the little *Microstylis*, one of the smallest and most inconspicuous of our orchids, while close beside it the colic-root raised its tall spike of white flowers and claimed recognition. Not far away from these, on a sandy knoll, one of the polygalas (*P. polygama*, Walt.) was quite abundant. At first sight it would not be claimed as of kin to the larger and better known fringed polygala, but a closer view shows it to be so. The same is true of another of the same genus (*P. sanguinea*, L.), in flower at the same time.

No fairer water scene can be found with us than the surface of a pond at early morning bedecked with the broad green leaves and the wide expanded flowers of the water-lily. There it reigns supreme amid the still life of the sequestered nook and in the busy life that swarms in the midsummer waters. It is one of our best known plants. It comes, too, of an ancient and honorable family. As Higginson says: "They assisted at the most momentous religious ceremonies, from the beginning of recorded time. The Egyptian Lotus was a sacred plant; it was dedicated to Harpocrates and to the god Nofr Atmoo,—*Nofr* meaning *good*, whence the name of our yellow lily, *Nuphar*. But the true Egyptian flower was *Nymphæa Lotus*. It was cultivated in tanks in the gardens; it was the chief material for festal wreaths; a single bud hung over the forehead of many a queenly dame; and the sculptures represent the weary flowers as dropping from the heated hands of belles, in the later hours of the feast. The Egyptian Lotus was, moreover, the emblem of the sacred Nile,— as the Hindoo species, of the sacred Ganges; and each was held the symbol of the creation of the world from the waters."

In the same waters are the numerous floating stems of several species of *Utricularia*, two of which, *U. vulgaris*, L. and *U. inflata*, Walt. are quite conspicuous with their yellow flowers. These are carnivorous plants, and woe to the little creature who enters the trap of the crowded little bladders borne on the leaves. On the sandy shore the round-leaved sundew and the long-leaved sundew may both be found, both carnivorous also, but naturally differing in habit from the bladderwort.

The two vervains are often as closely associated in their habitat as they are in the above list. The blue one was once held in high repute for its supposed medicinal virtues, but the wheel of its fortune has turned round, and the blue vervain is now a mere weed. The lopseed may be easily recognized even by the uninitiated, when its little purplish flowers are seen standing at right angles to the stem and, a little later, the ripening pods all pointing to the ground. We cannot miss seeing these tall stems with large leaves and yellow flowers, which prove to be elecampane.

The *Gaultheria* and the two species of *Chimaphila* cover the forest floor as we go in search of the ginseng, the rarest of our species of *Aralia*. Its habitat is in the cool dark woods, where on these hot July days we do not wonder it loves to dwell. There it is pleasant to seek for it, although the little brook which in spring flowed laughingly by is now hushed, and the pale mosses that lined its banks and covered the stones with such a soft green carpet now pine for its companionship again. There is now a hush upon the forest shades, but there is life there, rich and exuberant, full of change and full of charm.

# THE MIDSUMMER FLOWERS.

Down the glimmering lake there are miles of silence and still waters and green shores, overhung with a multitudinous and scattered fleet of purple and golden clouds, now furling their idle sails and drifting away into the vast harbor of the South.
— HIGGINSON — *Out-Door Papers.*

The traveler in the remote southwest of the United States, in that part originally settled by the Spaniards, cannot but appreciate the poetic impulse which led the early explorers of that region to call the rare springs in that arid land *ojos*, the Spanish for *eyes*. Ojo Oso, Ojo de Vaca, Ojo Caliente were and are the eyes of the landscape, giving it life and beauty; light shines out of them.

In our New England landscape the numerous ponds and lakes are as truly bright and flashing eyes, when seen from some hill or mountain-top, as any of the springs in the desert, but our familiarity with them may prevent us from appreciating their entire beauty. Yet we might pine for them, if we were far away, as sadly as Captain Hall's Esquimos pined to return to the ice of their old home.

The Autocrat of the Breakfast Table has experienced my pleasure in boating before me and has expressed it so fittingly that I will let him speak for me: "Here you are afloat with a body a rod and a half long, with arms or wings as you may choose to call them, stretching more than twenty feet from tip to tip; every volition of yours extending as perfectly into them as if your spinal cord ran down the center of your boat and the nerves of your arms tingled as far as the broad blades of your oars. This, in sober earnest, is the nearest approach to flying that man has ever made, or perhaps ever will make. I dare not publicly name the rare joys, the infinite delights, that intoxicate me on some sweet June morning, when the river and bay are smooth as a sheet of beryl-green silk and I run along ripping it up with my knife-edged shell of a boat, the rent closing after me like those wounds of angels which Milton tells of, but the seam still shining for many a long rood behind me."

A right fair sight is the lake which I know. The hills which rise from its margin, the woods, the dark shadows along the western shore in the afternoon, the bright light flooding the eastern shore, the slight, laughing ripple on the water, all combine to form as pretty a picture as I wish to see. How pleasant it was in those days when *The Sanctuary* was yet unviolated to glide from the broad bosom of the lake through the narrow channel winding hither and thither amid clumps of sweet gale and *Cassandra*, through patches of the yellow lily and over countless forms of aquatic life, into this sheltered nook. The wind was sighing in the tops of the tall pines on the left, and the light leaves of the chestnut were tossing all about on the right and in front, but no breath, at this depth, could ruffle the surface of the little pool. Save the great hawk which sailed away from among the pines as we entered, there was no sign of bird life. All was still as the shadows which at that hour lay around the margin.

We are tempted to look along the shore for those mild-eyed melancholy lotos-eaters bearing branches of that enchanted stem, laden with flower and fruit, which so charmed the companions of Ulysses; but after drinking in the charmèd beauty of the scene we paddle away again, watching the shores lined with shrubbery to the water's edge, with here and there a belated azalea covered with its sweet white blossoms, watching the water-grasses wave as we pass, until we emerge again over the sandy bar on the wider waters of the lake.

"And we followed the curving shore,
And ever northward bore."

On the western shore lies Wigwam Hill, from the top of which there is such a good view of the lake in its long extent. What a place for an arboretum of native trees and shrubs, many of which are already there! How the ferns love to dwell among the rocks on the sides of such a wooded height! Further on, under the bridge, we come into an expanded bay where the waters are shallower and the aquatic life more abundant. Here we find for the first time the white water crowfoot (*R. aquatilis*, L., var. *tricophyllus*, Gray), which we have in vain sought for elsewhere, and the shore is white with the blossoms of the interesting little pipewort. But the forms of life are almost bewildering in number; their variety fills us with a feeling of awe at the unbounded and still unknown resources of Nature. As we glide back along the shores traversed before, and watch the shadows lengthening over the water and the slanting sunlight gilding the hills and the clouds high over them, we see new objects of interest and realize more deeply than ever the world of beauty that lies close to our door, of which the following list forms a part.

July 24. 401 Elocharis ovata, R. Br.     Spike-Rush.
                Elymus Virginicus, L.           Wild Rye.
                Rhynchospora glomerata, Vahl    Beak-Rush.
                Laportea Canadensis, Gaudich.   Wood-Nettle.
                Hypericum mutilum, L.          St. John's-wort.

## THE MIDSUMMER FLOWERS.

|  |  | |
|---|---|---|
| | Dianthus Armeria, L. | Deptford Pink. |
| | Euphorbia maculata, L. | Spurge. |
| | Adiantum pedatum, L. | Maidenhair. |
| | Dicksonia pilosiuscula, Willd. | Dicksonia. |
| 410 | Asplenium Filix-foemina, Bernh. | Spleenwort. |
| July 27. | Eriocaulon septangulare, Withering | Pipewort. |
| | Cephalanthus occidentalis, L. | Button-bush. |
| | Dulichium spathaceum, Pers. | Dulichium. |
| | Tilia Americana, L. | Basswood. |
| | Phytolacca decandra, L. | Pigeon-Berry. |
| " 31. | Polygonum aviculare, L. | Knotweed. |
| | Apios tuberosa, Moench | Ground-nut. |
| | Desmodium Canadense, DC. | Tick-Trefoil. |
| 419 | Setaria viridis, Beauv. | Green Foxtail. |

| | | |
|---|---|---|
| LI | Rhododendron maximum, L. | Great Laurel. |
| | Monotropa uniflora, L. | Indian Pipe. |
| | Drosera intermedia, Hayne, var. Americana, DC. | Sundew. |
| | Brasenia peltata, Pursh | Water-Shield. |
| | Peltandra undulata, Raf. | Arrow Arum. |
| | Desmodium acuminatum, DC. | Tick-Trefoil. |
| | Rhexia Virginica, L. | Meadow-Beauty. |
| | Steironema lanceolatum, Gray | Loosestrife. |
| | Ilysanthes riparia, Raf. | False Pimpernel. |
| LX | Humulus Lupulus, L. | Common Hop. |
| | Goodyera repens, R. Br. | Rattlesnake-Plantain. |
| | Verbascum Blattaria, L. | Moth Mullein. |
| | Malva moschata, L. | Musk Mallow. |
| | Acorus Calamus, L. | Sweet Flag. |

|       |                                         |                          |
|-------|-----------------------------------------|--------------------------|
|       | Vitis cordifolia, Michx.                | Frost Grape.             |
|       | Ranunculus septentrionalis, Poir.       | Buttercup.               |
|       | Habenaria tridentata, Hook.             | Habenaria.               |
|       | Selaginella apus, Spring                | Dwarf Club-Moss.         |
|       | Veronica officinalis, L.                | Common Speedwell.        |
| LXX   | Euphorbia Preslii, Guss                 | Spurge.                  |
|       | Arabis Canadensis, L.                   | Sickle-pod.              |
|       | Arabis perfoliata, Lam.                 | Tower Mustard.           |
|       | Fumaria officinalis, L.                 | Fumitory.                |
|       | Geum strictum, Ait.                     | Avens.                   |
|       | Utricularia inflata, Walt.              | Bladderwort.             |
|       | Scleranthus annuus, L.                  | Knawel.                  |
|       | Ranunculus aquatilis, L., var. tricophyllus, Gray | White Water Crowfoot. |
|       | Rubus odoratus, L.                      | Purple Flowering Raspberry. |
|       | Onoclea Struthiopteris, Hoffm.          | Ostrich Fern.            |
| LXXX  | Geranium Robertianum, L.                | Herb Robert.             |
| 420   | Arctium Lappa, L., var. minus           | Burdock.                 |
|       | Vernonia Noveboracensis, Willd.         | Iron-weed.               |
|       | Convolvulus sepium, L.                  | Hedge Bindweed.          |
|       | Helianthus strumosus, L.                | Sunflower.               |
|       | Gerardia flava, L.                      | Downy False Foxglove.    |
|       | Panicum sanguinale, L.                  | Crab-Grass.              |
|       | Erigeron Canadensis, L.                 | Horse-weed.              |
|       | Prenanthes alba, L.                     | Rattlesnake-root.        |
|       | Alisma Plantago, L.                     | Water-Plantain.          |
|       | Hypericum nudicaule, Walt.              | Orange-grass.            |
| 430   | Eupatorium purpureum, L.                | Joe-Pye Weed.            |

## THE MIDSUMMER FLOWERS. 223

|  |  |
|---|---|
| Eupatorium perfoliatum, L. | Thoroughwort. |
| Tanacetum vulgare, L. | Tansy. |
| Panicum Crus-galli, L. | Barnyard-Grass. |
| Amarantus retroflexus, L. | Amaranth. |
| Amarantus albus, L. | Tumble Weed. |
| Rhus copallina, L. | Dwarf Sumach. |
| Polygonum Hydropiper, L. | Smartweed. |
| Phegopteris polypodioides, Fée | Beech Fern. |
| Aspidium cristatum, Swartz | Shield Fern. |
| 440 Aspidium cristatum, Swartz, var. Clintonianum | " " |
| Aug. 8. Aspidium Thelypteris, Swartz | " " |
| Aspidium Noveboracense, Swartz | " " |
| Cuscuta Gronovii, Willd. | Dodder. |
| Lobelia cardinalis, L. | Cardinal-flower. |
| Elodes campanulata, Pursh | Marsh St. John's-wort. |
| Lespedeza polystachya, Michx. | Bush-Clover. |
| Gnaphalium polycephalum, Michx. | Everlasting. |
| Mentha viridis, L. | Spearmint. |
| Andropogon scoparius, Michx. | Beard-Grass. |
| 450 Panicum agrostoides, Muhl. | Panic-Grass. |

Only one tree is found now in bloom, the basswood or linden or lime, as it is variously called. This and the chestnut are the last of our trees to put forth their blossoms. Of most of the others the fruits are already ripened. There are yet four or five interesting shrubs in bloom. The button-bush is easily recognized by its spherical-shaped heads of white flowers. It is confined to wet situations, frequently growing with its roots and part of the stem under water, forming large clusters on the edges of ponds or lining the banks of rivers. "It is associated," says Wilson Flagg, "with the complaining song of the blackbird, whose nest is often placed in the forks of its branches, and it accompanies the ruder aspects of nature."

The dwarf sumach, on the other hand, is found in dry places, lining on both sides the old Virginia rail fence or the stone wall of the upland pasture, although it does not despise the company of the merry brook. The dwarf sumach is easily distinguished from the other four species of *Rhus*, not so much by its size, as by the winged petiole of the leaf.

I saw the flowering raspberry (*Rubus odoratus*, L.) for the first time more than twenty years ago, in fruit on the slopes of Owl's Head by Lake Memphremagog; then in Princeton, and then close at home in Millbury and on Tataessit Hill in Worcester. It was the same experience to which I have alluded before; the remote was found to be near, when the eyes were opened to see it.

Rare and local, yet of great beauty, the *Rhododendron maximum*, L. of the Auburn swamp calls up pictures of many a pleasant pilgrimage to the spot where it deigns to abide. May it live there long and prosper till it fills the swamp with its beauty, and helps us to see with the mind's eye those gorgeous masses of color which cover the slopes of the southern Alleghanies and fill the higher valleys of the Himalayas!

The fruiting ferns are now abundant, forming eleven per cent. of this list. They thrive in the cool damp woods, though they do not dread the garish day. They are symbols of healthy strength and vigor, and form a rich setting for the flowering herbs near which they grow. Perhaps the favorite among them is the delicate maiden-hair, yet the beech-fern and the shield-fern and the ostrich-fern are equally worthy of our admiration.

In the dry sandy slope of the railroad embankment I have often found the wild rye-grass growing luxuriantly. Further south it frequently forms a considerable portion of native meadow lands and makes a coarse hay. It was counted by Dr. Vasey among the agricultural grasses of the United States, and it may be yet found worthy of cultivation. These little dark-brown, almost black, heads towering among the grasses of the lowland meadows are probably the flowering spikelets of the beak-rush, one of the great order Cyperaceæ, which in midsummer forms a large, if not conspicuous, part of the vegetation of this region. It is not difficult to recognize at once the relationship between the common nettle and the wood-nettle. A touch would prove it. In the shade by the roadside it grows tall and self-possessed. The spreading mats of the little spurge (*Euphorbia maculata*, L.) almost cover the space between the rails of the railroad and form great patches on the sandy shores of ponds. It basks in the open sunlight and thrives best in its smile.

The delicate flowers and the tender foliage of the early spring have now given place to the coarser herbage of summer. Tall stems rise up into the sight. Here is the pigeon-berry, not to be overlooked in flower, although it is more conspicuous when in fruit. Who does not know the burdock? Not to know it is to be ignorant of the boyish pranks for which it used to furnish material. Growing near it, and standing near it in the list, is a tall coarse plant, the iron-weed, the eastern representative of the genus Vernonia, more abundant further west. It gives a bit of bright color to the swampy meadow or the river-side where it is at home, and is most satisfactory, perhaps, when seen at a little distance. Some flowers are like oil-paintings in this respect: they produce the best effect when not examined too closely.

Compound flowers are now coming rapidly to the front, both in number of individuals and in number of species. These flowers by the roadside where unpruned Nature still holds sway are sunflowers, of which we have at least four species besides the garden one, which also sometimes escapes from cultivation. Besides these there are the tall-stemmed but inconspicuous-flowered horse-weed, the drooping-flowered *Prenanthes*, the Joe-Pye weed, the thoroughwort and the yellow tansy. These are eminently social flowers too: it is very rarely that a single plant of any of them is found growing by itself.

The faint sweet odor borne on the gentle breeze tells us that the brownish clusters of the ground-nut ought to be found near by. The downy fox-glove prepares the way for the other four members of its genus (*Gerardia*). Its yellow flowers are far-seen in the forest glades. The spearmint is a favorite mint even yet, and is often found in the neighborhood of neglected or abandoned farm-houses. No adept in wort-cunning can afford to be ignorant of its virtues. It was once regarded as the mint of mints—"the smell of which was believed to corroborate the brain and increase and preserve the memory, and it was venerated like one of the holy herbs." The day of simples and of simplers is nearly gone, and with these are going those old superstitions which spread the charm of romance over a great part of the vegetable kingdom.

> "'The old men studied magic in the flowers,
> And human fortunes in astronomy,
> And an omnipotence in chemistry,
> And, wheresoever their clear eye-beams fell,
> They caught the footsteps of the SAME."

# THE MID-AUGUST FLOWERS.

> Along the roadside, like the flowers of gold
> That tawny Incas for their gardens wrought,
> Heavy with sunshine droops the golden-rod,
> And the red pennons of the cardinal flowers
> Hang motionless upon their upright staves.
> The sky is hot and hazy, and the wind,
> Wing-weary with its long flight from the south.
> — WHITTIER — *Among the Hills.*

In these mid-August days the doors of summer are open wide. The tide of life is not far from the full. The time is propitious to explore the southern end of the lake, and we find ourselves gliding along among the islands, and skirting the city's new Lake Park, admiring its situation and looking forward a few years to the days when here Nature and Art joining hand to hand shall make a perfect union. Half-Moon Pond does not long detain us, and a portage is soon made over the highway which forms its southern boundary, for the waters of Flint's Pond invite us. And now we are in a shallow pool whose surface is strewn with many forms of plant life.

The emerald fringe of the margin shuts out any distant prospect. Above us is the sky, flecked with gray clouds, around us are the water and the woods, bathed with ever-changing lights and shadows, the whole as calm and peaceful as a picture out of fairyland. Lightly we row along and slowly, gathering the fragrant lilies along our path, pushing our boat into this quiet nook and into that in search of more of these beauties. Here our progress is hindered by the sight of such a group of cat-tail flags (*Typha latifolia*, L.) as we cannot resist the gathering, though they are sheltered behind a strong defense of the poison dog-wood; and now we glide slowly across the broad open through the shield-shaped leaves of the Brasenia, clutching now this, now that little Potamogeton or trailing stem of Utricularia, around whose slender leaves we know the compound microscope will reveal a whole other world of rare forms and entrancing beauties, or along the delightfully wooded shore where the chestnut and the oak, the birch and the maple fill the higher places, and the button-bush and the *Clethra*, both in bloom, line the margin.

Yonder to the right is the great dam, beyond which these waters will be harnessed to wheel after wheel as they go hurrying onward to the ocean. Their childhood days are now ended, their busy days are come. As we turn homeward, choosing the shorter way, beneath the bridges, we leave the long pastures sloping down to the water, leave the fields of corn waving in the gentle breeze and the farm-houses scattered on the hillsides, and soon we are again in the wilds. As we round a corner quickly we startle two bitterns, which fly into the dense shrubbery of the further shore and are quickly lost to our sight. We feel almost guilty for intruding into such a quiet nook. There is such an air of privacy about it, the public can have scarcely any rights here.

It will be pleasant to stay a little on the edge of this open glade and watch the life around us. From a tree projecting over the water a little ahead of us come the sharp notes of a kingfisher; my companion's quick ear has caught the fainter note of the wood-pewee, and he points out to me a black and white creeper on a tree not far away. A summer warbler flying close by alights in the nearest shrub, and a red squirrel is disporting himself yonder, rushing up and down as though his life depended upon it. Our attention is attracted by a splash in the water, and we see the kingfisher rising to his perch again. Apparently he was unsuccessful, for, see! there he goes again, straight as an arrow and plunges beneath the water. This time we think he has succeeded, for we see him no more, yet we hear a sort of chuckle which indicates satisfaction. The sun is now shining athwart the surface, so that in these shallows our kingfisher from his high perch could take in a clear view of the waters.

After a time we push out from our cosy nook toward a dam which separates us from the broader bosom of the lake. But as we go, we cannot enough admire those reaches of water over which we have come, nor the wonderful play of light in the distant coves, nor the deepening shadows under the trees, nor the quiet brooding on the Shrewsbury hills yonder where the solitary elms rear their vase-like forms against the sky. Before us the level sun makes a broad track of light, like that along which Sir Bedivere saw the royal barge pass with the wounded King Arthur. And ever as we pass some new point, a new vista opens behind us as well as before. Soon we are again among the islands, and are skirting the shore already traveled over, the same shore and yet different, as it lies now enveloped in shadow. And the water is smooth, and the moon is up, and the silence of a summer night is settling over the lake. Soon the stars will be flashing out overhead, from below other stars will answer to them. We have seen something of the charms of our favorite lake, yet we know we have not seen them all.

What Starr King says of Lake Winnipesaukee, we may say of Lake Quinsigamond: "Seen the lake! Which lake? There are a thousand. It is a chameleon. It is not a steady sapphire set in green, but an opal.

Under no two skies or winds is it the same. It is gray, it is blue, it is olive, it is azure, it is purple, at the will of the breezes, the clouds, the hours. Sail over it on some afternoon when the sky is leaden with northeast mists, and you can see the simple beauty of form in which its shores and guards are sculptured."

We must see it when clouds are flying over the sun, when the sky is cloudless, when a thunder-storm is sweeping over it, when the breezes ripple its surface, when it lies smooth as ice, when the morning sun is breaking over it, when the setting sun tinges the hills all about it with rich purple light, when spring is clothing its margin with the first faint touches of green, and when autumn clothes it with the gorgeous hues which make our forests glad, and even later when the leaves have flown from all but the stately pines and the waters are quiet beneath their robe of ice and snow.

Those who love it best will see most of its beauties, and those who visit it oftenest will love it best. We bring away from it a folio of sunny memories. The city streets seem all the better and purer for the few hours spent

> "About the windings of the marge to hear
> The soft wind blowing over meadowy holms
> And alders, garden isles; where now we ran
> By ripply shallows of the lisping lake,
> Delighted with the freshness and the sense
> Of restfulness."

## THE MID-AUGUST FLOWERS. 233

The following list is typical of the abundance of plant-life in these mid-August days. A glance backward will show that the first part of August 8th belongs to the preceding list; the last part of August 17th belongs to the succeeding one: so overflowing are the fountains of Life at this festal season. It will be borne in mind, too, that if I should count all those found in bloom on these two days, many of which are recorded in the preceding lists, the number would be largely increased.

| | | | |
|---|---|---|---|
| Aug. 8. | 451 | Solidago nemoralis, Ait. | Golden-rod. |
| | | Cyperus strigosus, L. | Galingale. |
| | | Hypericum Canadense, L. | St. John's-wort. |
| | | Lobelia inflata, L. | Indian Tobacco. |
| | | Scutellaria lateriflora, L. | Mad-dog Scullcap. |
| | | Anaphalis margaritacea, Benth. & Hook. | |
| | | | Pearly Everlasting. |
| | | Clethra alnifolia, L. | Sweet Pepperbush. |
| | | Ambrosia artemisiæfolia, L. | Roman Wormwood. |
| | | Gratiola aurea, Muhl. | Hedge-Hyssop. |
| | 460 | Aster umbellatus, Mill. | Aster. |
| | | Helianthus divaricatus, L. | Sunflower. |
| | | Epilobium lineare, Muhl. | Willow-herb. |
| | | Solidago lanceolata, L. | Golden-rod. |
| " 17. | | Spiranthes gracilis, Bigelow | Ladies' Tresses. |
| | | Sagittaria variabilis, Engelm. | Arrow-head. |

| | |
|---|---|
| Goodyera pubescens, R. Br. | Rattlesnake-Plantain. |
| Penthorum sedoides, L. | Ditch Stone-crop. |
| Acalypha Virginica, L. | Three-seeded Mercury. |
| Lespedeza capitata, Michx. | Bush-Clover. |
| 470 Hedeoma pulegioides, Pers. | Pennyroyal. |
| Desmodium ciliare, DC. | Tick-Trefoil. |
| Hieracium paniculatum, L. | Hawkweed. |
| Hieracium scabrum, Michx. | " |
| Gerardia pedicularia, L. | Foxglove |
| Gerardia purpurea, L. | Purple Gerardia. |
| Polygonum arifolium, L. | Halberd-leaved Tear-thumb. |
| Trichostema dichotomum, L. | Blue-Curls. |
| Amphicarpæa monoica, Nutt. | Hog Peanut. |
| Aralia racemosa, L. | Spikenard. |
| 480 Cenchrus tribuloides, L. | Bur-Grass. |
| Panicum capillare, L. | Old-witch Grass. |
| Lactuca leucophæa, Gray | False Lettuce. |
| Epilobium coloratum, Muhl. | Willow-herb. |
| Solidago rugosa, Mill. | Golden-rod. |
| Paspalum setaceum, Michx. | Paspalum. |
| Muhlenbergia Wildenovii, Trin. | Drop-seed-Grass. |
| Aster cordifolius, L. | Aster. |
| Aster puniceus, L. | " |
| Chelone glabra, L. | Snake-head. |
| 490 Habenaria psycodes, Gray | Fringed Orchis. |
| Sedum Telephium, L. | Live-for-ever. |

THE MID-AUGUST FLOWERS. 235

    Sparganium eurycarpum, Engelm.   Bur-reed.
    Gnaphalium uliginosum, L.   Low Cudweed.
    Asplenium ebeneum, Ait.   Spleenwort.
    Asplenium Trichomanes, L.   "
    Asplenium thelypteroides, Michx.   "
    Onoclea sensibilis, L.   Sensitive Fern.
    Polypodium vulgare, L.   Polypody.
    Lycopodium clavatum, L.   Common Club-Moss.
500 Lycopodium obscurum, L., var. dendroideum, Michx.   Ground-Pine.

The trees have now put forth their blossoms; only one or two shrubs remain to grace the end of the year, but among them is the *Clethra*. The margin of the little pool, where in the rare days of early June the pink azalea filled the air with its fragrance, and where, a little later, the white racemes of the *Leucothoe* smiled amid the bright spring verdure, is now lighted up by the far-seen snowy *Clethra*, while the sweet southwest wind playing over the water wafts its faint odor far. Fortunate little Pool, basking through the summer months in the smiles of three so fair beauties! It is well for thee that thou hast no golden apple to bestow upon the fairest, so to win the hatred of the other two and see them turn their faces away from thy banks. Attractive as the *Clethra* is in itself, it is still more attractive from the fact of its being almost alone among the shrubs in its time of flowering. It well repays cultivation by the increase in the size and the number of its flowers.

What more fitting leader could this list have than one of the golden-rods! It may be dusty from growing by the roadside, and may be despised as a weed, but its kindred will yet brighten all the late summer fields with their abundance of yellow.

Five species of *Hypericum* have already come and gone, or are still lingering as if waiting for their youngest species, *Canadense*. Close behind is a *Lobelia*, the species *inflata*, the last of its genus, the delicate *spicata* and the bold *cardinalis* having just preceded it. The little skull cap is seen where marsh plants abound, and is easily recognized by its small blue flowers arranged in pairs along the stem. Where the pond has retreated from the shore, leaving to the sun to warm and vivify the muddy margin, the little hedge-hyssop may often be found lifting its golden-yellow head among the grasses and sedges there disputing the supremacy.

The orchid family has been well represented hitherto, and sends three representatives to the present list. They are of no mean rank, the delicate ladies' tresses, the more showy *Goodyera* with its white-streaked leaves nestling in clusters close to the ground yet raising a spike of white flowers to please the eyes of the finder, and the smaller purple-fringed orchis, which, when once seen, is not easily forgotten.

Those arrow-headed leaves in the pool suggest the *Sagittaria* very plainly, and close by it I may expect to find the stone-crop. In the pasture I must have unwittingly trodden upon the pennyroyal, for its odor fills the air. That — bush I had almost called it — plant covered with large yellow flowers and buds, with cut leaves, is a *Gerardia*, and this little plant in the moist roadside or low-lying meadow, with rose-purple flowers of the same somewhat tubular shape, is another of the family. I shall not have to travel far along the railroad embankment at this season to find blue-curls and, probably that outcast among the grasses, *Cenchrus tribuloides*. In the untrimmed vegetation of the roadside I look now for the last *Aralia*. Early in May I found *A. trifolia;* at the end of the month came *A. nudicaulis*, extending its presence into June; as it passed away *A. hispida* appeared, followed closely by *A. quinquefolia;* now *A. racemosa* comes, the last of its family.

A quarter of the plants of this list belong to the great order *Compositæ*. The leaders of the aster family are at hand. Their broad banner of purple and white will be fully unfolded by and by. The snake-head or turtle-head lifts up into the sunlight its spike of white flowers, from among the grasses and sedges and goldenrods of the swamps. The twining stems of the *Amphicarpæa*, with its purplish nodding racemes, help to fill up any gaps in the foliage along the margin of the woodland. This plant has many showy flowers but they seldom ripen any fruit, and some rudimentary and inconspicuous flowers, which are generally fruitful. In this respect it resembles the violets.

The ferns, too, are now very abundant; none of them is more attractive to me than the graceful *Asplenium ebeneum*. I must often examine under the compound microscope these brown spots upon the back of the frond and try to realize, if only imperfectly, the marvelous contrivances and inexhaustible variety shown in the vegetable world. Who knows but that there are other worlds of vegetable forms as far beyond the power of the microscope to discover as that which it discovers to us is beyond the power of our unaided vision? Here, in the *Onoclea*, we find at first only sterile fronds with no sign of fruitage on the back in the shape of either rounded or oblong spots, or coiled up under the margin as in the maiden-hair. But if we push these sterile fronds aside we shall find here and there among them the shorter stems, surmounted by greenish clusters resembling small berries, which are the fertile fronds and do not suggest ferns.

The order *Lycopodiaceæ*, like the ferns, stands near the head of the flowerless plants. Like them, too, it has a remote ancestry. In the forests of the Carboniferous Age gigantic members of this family flourished, whose remains are found in great abundance in the coal strata. In the changed conditions of life they now play an unimportant part. My earliest recollection of them is associated with Christmas festivities, where in festoon and wreath they formed the larger part of the decoration of the village church, a service which, in New England at least, they seem destined for some time to fulfil. Persistent survivors of a declining family, they tell an interesting chapter of the earth's history.

# THE LATE SUMMER FLOWERS.

> I thought I knew all Summer knows,
> So many summers I had been
> Wed to Summer. Could I suppose
> One hidden beauty still lurked in
> Her days? that she might still disclose
> New secrets and new homage win?
> — HELEN JACKSON — *This Summer.*

How imperceptibly the late summer glides into the early autumn as if to lessen our feeling of regret at its passing with all the pleasures of rambles through mead and glade! As the flame often flashes with unusual brilliancy just before expiring, so the earth clothes itself with gayest robes before preparing for its winter sleep. There are no gloomy hues in the late summer flowers. The season is one of brightness, of calm, of halcyon days. There is no sense of hurry; rather a feeling that there will be time for every little duty to be accomplished.

There is a myth quite prevalent in New England, almost coeval with the early settlement of the country, that somewhere in the autumn, after the leaves have fallen, there comes a delightful time, longer or shorter, when

> "Falls not hail or rain or any snow,
> Nor ever wind blows loudly,"

which is known as the Indian Summer.

The Indian and the primeval forest through which he roamed on his hunting expeditions have both disappeared; and with them has gone, blown away like them before the chilling breath of civilization, and like them nevermore to return, that charming season upon the beauties of which so many poets and prose writers have loved to dwell. It was a belief of the Indian that the Great Spirit sent this mild season in the late autumn for the especial benefit of his red children; and when the great fall hunt was over and the floor of the wigwam was covered with the spoils of the forest chase, a season of festivity followed suggestive of modern harvest festivals.

But there is no myth about the beauty of these late summer days. The glory of the year is wrapped up in them. They call to mind the picture of the fair-tressed Pallas Athené as she

> "Rose, like a pillar of tall white cloud, toward silver
> Olympus:

> Far above ocean and shore, and the peaks of the
>     isles and the mainland;
> Where no frost nor storm is, in clear blue windless
>     abysses,
> High in the home of the summer, the seats of the
>     happy Immortals,
> Shrouded in keen deep blaze, unapproachable."

They are full of fulfilment. The promise of other days is realized. They invite to rest after the burden and heat of the day. No other part of the year is just like them. It is always true that

> "The old order changeth, yielding place to new,
> And God fulfils himself in many ways,
> Lest one good custom should corrupt the world."

We have waited longer than usual for our last group of flowers, but they make a brave show when they appear.

| | | | |
|---|---|---|---|
| Aug. 17. | 501 | Gerardia quercifolia, Pursh | Smooth False Foxglove. |
| | | Echinospermum Virginicum, Lehm. | Beggar's Lice. |
| | | Aster Tradescanti, L. | Aster. |
| | | Pilea pumila, Gray | Richweed. |
| " | 29. | Decodon verticillatus, Ell. | Swamp Loosestrife. |
| | | Cyperus dentatus, Torr. | Galingale. |
| | | Aster patens, Ait. | Aster. |

## THE LATE SUMMER FLOWERS. 243

Aug. 31.
        Aster Novæ-Angliæ, L.     Aster.
        Aster Novi-Belgii, L.       "
 510 Solidago bicolor, L.     White Golden-rod.
        Sicyos angulatus, L.      One-seeded Bur-Cucumber.
        Collinsonia Canadensis, L.   Stone-root.
        Eupatorium ageratoides, L.  White Snake-root.
 514 Leersia oryzoides, Swartz   Rice Cut-Grass.
LXXXI Desmodium nudiflorum, DC.   Tick-Trefoil.
        Desmodium paniculatum, DC.  "  "
        Spiranthes præcox, Watson   Ladies' Tresses.
        Solidago serotina, Ait.      Golden-rod.
        Xyris flexuosa, Muhl., var. pusilla, Gray
                              Yellow-eyed Grass.
        Desmodium rigidum, DC.    Tick-Trefoil.
        Desmodium Marilandicum, F. Boott  "  "
        Panicum glabrum, Gaudin   Panic-Grass.
        Trifolium procumbens, L.    Low Hop-Clover.
  xc Cnicus altissimus, Willd., var. discolor, Gray
                              Thistle.
        Ludwigia palustris, Ell.     Water Purslane.
        Cinna arundinacea, L.      Wood Reed-Grass.
        Chrysopogon nutans, Benth.   Indian Grass.
        Rosa Carolina, L.         Carolina Rose.
        Sium Carsonii, Durand      Water Parsnip.
        Lespedeza Stuvei, Nutt.     Bush-Clover.
        Lespedeza procumbens, Michx.  "  "
        Cystopteris fragilis, Bernh.   Bladder Fern.
        Pycnanthemum lanceolatum, Pursh    Basil.

|   |   |   | | |
|---|---|---|---|---|
| | | c | Pycnanthemum linifolium, Pursh | Basil. |
| | | | Bidens Beckii, Torr. | Water Marigold. |
| | | | Heliopsis lævis, Pers. | Ox-eye. |
| | | cIII | Phalaris Canariensis, L. | Canary Grass. |
| Sept. | 4. | 515 | Bidens frondosa, L. | Common Beggar-ticks. |
| | | | Aster linariifolius, L. | Aster. |
| | | | Andropogon furcatus, Muhl. | Beard-Grass. |
| | | | Aster dumosus, L. | Aster. |
| " | 5. | | Bœhmeria cylindrica, Willd. | False Nettle. |
| | | 520 | Monotropa Hypopitys, L. | Pine-sap. |
| | | | Corallorhiza multiflora, Nutt. | Coral-root. |
| | | | Desmodium rotundifolium, DC. | Tick-Trefoil. |
| | | | Solidago Canadensis, L. | Golden-rod. |
| | | | Bidens connata, Muhl. | Swamp Beggar-ticks. |
| | | | Muhlenbergia sylvatica, Torr. & Gray | |
| | | | | Drop-seed Grass. |
| | | | Aster corymbosus, Ait. | Aster. |
| " | 12. | | Aster multiflorus, Ait. | " |
| | | | Aster lævis, L. | " |
| | | | Aster salicifolius, Ait. | " |
| | | 530 | Aster undulatus, L. | " |
| | | | Parnassia Caroliniana, Michx. | Grass of Parnassus. |
| " | 16. | | Proserpinaca palustris, L. | Mermaid-weed. |
| | | | Spiranthes cernua, Richard | Ladies' Tresses. |
| | | | Gentiana Andrewsii, Griseb. | Closed Gentian. |
| | | | Bidens chrysanthemoides, Michx. | Larger Bur-Marigold. |
| | | | Lespedeza violacea, Pers. | Bush-Clover. |

## THE LATE SUMMER FLOWERS. 245

|  |  |  |
|---|---|---|
|  | Solidago odora, Ait. | Sweet Golden-rod. |
|  | Solidago cæsia, L. | Golden-rod. |
|  | Aster acuminatus, Michx. | Aster. |
| 540 | Aster macrophyllus, L. | " |
|  | Lycopus Virginicus, L. | Bugle-weed. |
|  | Solidago serotina, Ait., var. gigantea, Gray | |
|  |  | Golden-rod. |
|  | Lycopodium complanatum, L. | Ground-Pine. |
|  | Lycopodium lucidulum, Michx. | Stag-horn Moss. |
|  | Pteris aquilina, L. | Common Brake. |
| Sept. 19. | Mentha piperita, L. | Peppermint. |
|  | Solidago latifolia, L. | Golden-rod. |
|  | Helianthus tuberosus, L. | Jerusalem Artichoke. |
|  | Solidago Elliottii, Torr. & Gray | Golden-rod. |
| 550 | Hieracium Canadense, Michx. | Hawkweed. |
| CIV | Datura Stramonium, L. | Thorn-Apple. |
|  | Datura Tatula, L. | Purple Thorn-Apple. |
|  | Geranium Carolinianum, L. | Cranesbill. |
|  | Physalis Virginiana, Mill. | Ground Cherry. |
|  | Prenanthes altissima, L. | Tall White Lettuce. |
|  | Sporobolus vaginæflorus, Vasey | Drop-seed Grass. |
| CX | Sonchus oleraceus, L. | Sow-Thistle. |
|  | Polygala verticillata, L. | Polygala. |
|  | Medicago denticulata, Willd. | Medick. |
|  | Lygodium palmatum, Swartz | Climbing Fern. |
|  | Eragrostis major, Host | Eragrostis. |
|  | Erechtites hieracifolia, Raf. | Fireweed. |

Polygonum amphibium, L.     Knotweed.
Polygonum Muhlenbergii, Watson     "
CXVIII Galinsoga parviflora, Cav.     Galinsoga.

The Rear of the Procession.

Oct. 2. 551 Botrychium ternatum, Swartz, var. intermedium,     Moonwort.
Gentiana crinita, Frœl.     Fringed Gentian.
553 Hamamelis Virginiana, L.     Witch-Hazel.

The leader of this band is the oak-leaved Gerardia, gay with large yellow flowers. If we mark them at all, we see that yellow is a favorite color, for here are at least seventeen which display it. Nearly as many are robed in white, and not a few are clad in purple or blue.

What an admirable provision is this by which, willingly or unwillingly, we are made to assist *Echinospermum* and the various species of *Bidens* in scattering their seeds, as we pick from our clothing the little burs which attached themselves to us as we brushed by them! The almost transparent stems of the richweed furnish excellent material for microscopic study. It is abundant in low rich lands, especially in the neighborhood of barns, where it sometimes crowds out nearly every other form of vegetation. Its kinship to the nettle is easily recognized on anything more than the most superficial observation.

I first found the swamp loosestrife on the margin of a floating isle, where its long recurved stems, covered with axillary, rose-purple flowers, could not fail to attract the attention. I have found it since in other places but similar in situation. The shore of the mainland near by is thickly lined with *Cyperus dentatus*, one of the most showy of our galingales.

The asters and the golden-rods are everywhere. The late summer would scarcely seem to be itself without them. White, purple and yellow, they fill the world with a beauty unbought but not unprized. Along the river-banks *Sicyos* is found abundant. The climbers have one advantage over other herbaceous plants in that they can cover so much space that they are not overlooked, and they can get up into the light. *Sicyos* is almost our only uncultivated species of the gourd family, represented in cultivation by the pumpkin, squash, cucumber, watermelon, etc.

This dried specimen of *Leersia* in my herbarium now calls to mind a late August day several years ago when its odd-looking panicles first caught my eye as I strolled along familiar paths and recognized a grass not seen before. Late summer has its favorite and peculiar grasses, as well as early spring. Most of them stand up tall and graceful with large panicles of purple, and yellow, and greenish flowers. *Cinna arundinacea* and *Chrysopogon nutans* are two of the most characteristic of these. Where the woods have been recently cut and the sun has had an opportunity to reach the earth with its warmth and vigor, these grasses and similar ones are most commonly found. The species of *Andropogon* display their feathery stigmas over many a New England pasture every autumn and persist until the coming of the snow. Perhaps they are, after all, the most characteristic of our late summer grasses, principally from their abundance.

Our several species of *Desmodium* will not be overlooked. If we do not find them all in one summer, we shall probably see them all in the course of two or three seasons. A few of them flourish by the roadside, but mostly they prefer the cool and shaded woodland. The same is true of the related genus *Lespedeza*, several of the species of which we cannot fail to notice at this time of year, as they crowd together along the wayside.

The little yellow heads of *Xyris* are so small that they may be overlooked, but they will repay some attention. They may be found on the gravelly shores of our numerous ponds. While one species of water parsnip is found throughout North America, the other, *Sium Carsonii*, is credited to only four states. I found it several years since in a cranberry bog in Sutton, which is as yet the only locality in which I can look for it. In itself it is not especially interesting, as few of its order, the Umbelliferæ, are.

The ferns have now mostly passed their prime, but there are two which have still a special interest. It was a surprise and a pleasure to me to find the bladder fern, (*Crystopteris fragilis*, Bernh.), in the woods along Lake Quinsigamond, after having looked long and patiently for it elsewhere. To me it is one of the most graceful and delicate of our native ferns, rivaled only by a few, among which I must admit the climbing fern (*Lygodium palmatum*, Swartz). This is perhaps the rarest of our local ferns, certainly in this part of our county. It is the only one of them with the power of climbing; and when growing in profusion it covers the neighboring shrubbery with a tangle of the most delicate greenery. I have in mind now a meadow a dozen miles away, with a broad clear brook running through it, and a wilderness of shrubbery enclosing it, where, on an early October day some years ago, I gathered it for the first time. My own experience enables me to appreciate the enthusiasm of the fern-collectors who have ransacked all parts of the earth for these beautiful forms.

When I thought I had found all the species of *Bidens* that probably occur in this locality, one more species (*B. Beckii*, Torr.), the water marigold, presented itself to me at the upper end of Lake Quinsigamond, near where the water crowfoot had previously been found. This gives us five of the six species of *Bidens* as described in Gray's Manual. It would not be at all surprising if the remaining species, which is found in Rhode Island and southward, should soon be found here also.

*Heliopsis lævis*, Pers. has come hither from further west, presumably in grass-seed, but has evidently come to stay, like *Rudbeckia* and others of its kin. *Galinsoga parviflora*, Cav. is an immigrant recently arrived from tropical America. Its coming into various parts of the country has been duly noted in the botanical journals. It is a weed, generally found about gardens now in the northern states. I made my first acquaintance with it one October, when I found it growing in the middle of a path through woods at least half a mile from any house and in a very quiet and secluded spot. I had often passed over that path in former years at all times of the year, but had not before detected the presence of *Galinsoga*. It was evidently a new arrival and, though a native of the sunny South, has been able to adapt itself in the struggle for existence to the changed conditions.

*Parnassia* and *Spiranthes* and *Gentiana* are three of the showy and attractive flowers of this time. There is a delicacy about the first two which makes them much sought after when once known. The closed flowers of the latter, bronzy-blue, have a special interest in connection with the subject of cross- and self-fertilization in the vegetable kingdom. Several species of *Lycopodium* are now in their prime, and the brake is ripening its spores and scattering them as widely as it can. Here and there a late hawkweed shows its yellow flowers, and the artichoke run wild lifts its sunflower-head in the copse. Thorn-apples, coarse and rank, have taken possession of neglected gardens in rivalry of the sow-thistle and its kindred.

When the fringed gentian and the witch-hazel come we know that the summer is passed. The leaves have fallen or are falling. The naked branches of the witch-hazel, adorned with the thickly-clustered flowers of yellow strap-shaped petals, present a weird appearance. It, too, is hopeful of another year, of another returning sun to complete the work which is this season left unfinished. It looks through the dark abyss of winter to the warm sunny skies beyond. Shall we be less brave or less hopeful?

APPENDIX A.

# FLORA OF WORCESTER COUNTY.

## A CATALOGUE

OF THE

PHÆNOGAMOUS AND VASCULAR CRYPTOGAMOUS

PLANTS OF WORCESTER COUNTY,

MASSACHUSETTS.

SECOND EDITION, REVISED AND ENLARGED.

1894.

## PREFACE.

In the autumn of 1883 I prepared for publication the first *Catalogue of the Plants of Worcester County, Massachusetts*, containing eight hundred and twelve species and well-marked varieties. In the meantime, by the kind and generous help of many friends, all of whom will understand, I hope, that whether their names appear in these pages or not, my sense of obligation to them is here fully acknowledged and gratefully remembered, more than two hundred additions have been made.

This Catalogue, revised and enlarged, is now issued in the hope that it may serve as a basis for still further additions, so that, in due time, our county may have its flora as fully recorded as possible. Its main purpose is to help and to encourage the beginner and to afford pleasure to those who take an interest in untamed and unpruned Nature.

Much yet remains to be done. The mosses, lichens, fungi and algæ furnish a wide field for the coöperation of workers in different parts of the county for some years. The yearly increasing interest in such work and the greatly increased facilities for prosecuting it are hopeful signs of its accomplishment.

The arrangement and the nomenclature of this Catalogue are those of Gray's *Manual of the Botany of the Northern United States,*—Sixth edition, 1890—the book most readily accessible to all persons interested in the study of our local flora. In the present state of botanical nomenclature, and for the purpose for which this Catalogue is intended, I have not thought it wise or necessary to adopt proposed changes not yet generally accepted. The student interested in such matters can easily adapt himself to changing conditions.

Introduced species, so far as I have been able to determine the fact of introduction, are indicated by a *.

# FLORA OF WORCESTER COUNTY.

## Series I.  PHÆNOGAMIA.
### Class I.  DICOTYLEDONEÆ.
#### Sub-class I.  ANGIOSPERMÆ.
##### Division I.  POLYPETALÆ.

### RANUNCULACEÆ.

CLEMATIS, L.    Virgin's Bower.
    Virginiana, L.                                                Common.
ANEMONE, Tourn.    Anemone.
    cylindrica, Gray.    Long-fruited Anemone.
    Virginiana, L.                                                Common.
    nemorosa, L.    Anemone.                       Very common.
HEPATICA, Dill.    Liver-leaf.  Hepatica.
    triloba, Chaix.                                               Common.
ANEMONELLA, Spach.
    thalictroides, Spach.    Rue-Anemone.           Not rare.

THALICTRUM, Tourn.     Meadow-Rue.
    dioicum, L.     Early Meadow-Rue.     Rocky woods.
    polygamum, Muhl.    Tall Meadow-Rue.     Common.
    purpurascens, L.    Purplish Meadow-Rue.     Grafton.
RANUNCULUS, Tourn.     Crowfoot.    Buttercup.
    aquatilis, L., var. trichophyllus, Gray.    White Water-
        Crowfoot.                             Worcester, Gardner.
    multifidus, Pursh.    Yellow Water-Crowfoot.
    Flammula, L., var. reptans, E. Meyer.     Lancaster.
                                                       *F. L. Palmer.*
    abortivus, L.    Small-flowered Crowfoot.     Common.
    recurvatus, Poir.    Hooked Crowfoot.    Woods, common.
    fascicularis, Muhl.    Early Buttercup.         Dry hills.
       "          "         var.    Sunnyside, Worcester.
                                                         *Prof. Eaton.*
    septentrionalis, Poir.                         Worcester.
    Pennsylvanicus, L. f.    Bristly Buttercup.    Barre.
                                                    *Miss Sara Lane.*
    *bulbosus, L.    Bulbous Buttercup.         Common.
    *acris, L.    Tall Buttercup.            Very common.
CALTHA, L.     Marsh Marigold.
    palustris, L.    Cowslip, but not of literature.    Common.
COPTIS, Salisb.     Goldthread.
    trifolia, Salisb.    Three-leaved Goldthread.    Common.
AQUILEGIA, Tourn.     Columbine.
    Canadensis, L.    Wild Columbine.            Common.
ACTÆA, L.     Baneberry.
    spicata, L., var. rubra, Ait.    Red Baneberry.    In copses.
    alba, Bigel.    White Baneberry.            With the last.

## MAGNOLIACEÆ.

LIRIODENDRON, L.   Tulip-Tree.
  Tulipifera, L.                              Rare.  *G. E. Stone.*

## BERBERIDACEÆ.

BERBERIS, L.   Barberry.
  *vulgaris, L.  In eastern and southern parts of the county.
CAULOPHYLLUM, Michx.   Blue Cohosh.
  thalictroides, Michx.   Pappoose-root.   Worcester, Princeton.
PODOPHYLLUM, L.   May-Apple.  Mandrake.
  peltatum, L.              West Boylston.  *Arba Pierce.*

## NYMPHÆACEÆ.

BRASENIA, Schreber.   Water-Shield.
  peltata, Pursh.                             Common.
NYMPHÆA, Tourn.   Water-Lily.
  odorata, Ait.  Sweet-scented Water-Lily.   Common.
  "      "      var. minor, Sims.             Rare.
NUPHAR, Smith.   Yellow Pond-Lily.
  advena, Ait. f.                             Common.
  Kalmianum, Ait.    Southbridge.  *L. E. Ammidown.*

## SARRACENIACEÆ.

SARRACENIA, Tourn.   Side-saddle Flower.
  purpurea, L.  Pitcher-Plant.                Common.

## PAPAVERACEÆ.

SANGUINARIA, Dill.   Blood-root.
  Canadensis, L.          Common on rocky hillsides.

CHELIDONIUM, L.   Celandine.
  *majus, L.   Celandine.   Waste grounds.
PAPAVER, Tourn.   Poppy.
  *somniferum, L.   Common Poppy.   Sutton.
*Dr. G. C. Webber.*

## FUMARIACEÆ.

ADLUMIA, Raf.   Climbing Fumitory.
  cirrhosa, Raf.   . Rare.
DICENTRA, Borkh.   Dutchman's Breeches.
  Cucullaria, DC.   Dutchman's Breeches.   Barre.
*Miss Sara Lane.*
  Canadensis, DC.   Squirrel Corn.   Barre, Winchendon.
CORYDALIS, Vent.   Corydalis.
  glauca, Pursh.   Pale Corydalis.   In dry woods.
  aurea, Willd.   Golden Corydalis.   Winchendon.
*F. R. Hathaway.*
FUMARIA, Tourn.   Fumitory.
  *officinalis, L.   Common Fumitory.   Waste places.

## CRUCIFERÆ.

DENTARIA, Tourn.   Pepper-root.
  diphylla, L.   North Brookfield.   *Miss E. M. Reed.*
CARDAMINE, Tourn.   Bitter Cress.
  rhomboidea, DC.   Spring Cress.   Wet meadows.
  hirsuta, L.   Small Bitter Cress.   Wet places.
  parviflora.   A form answering to this is found on Mt. Wachusett.
ARABIS, L.   Rock Cress.
  Canadensis, L.   Sickle-pod.   Sutton.
  perfoliata, Lam.   Tower Mustard.   Sutton, Worcester.

ALYSSUM, Tourn.     Alyssum.
    *calycinum, L.     Occasional.
CAMELINA, Crantz.     False Flax.
    *sativa, Crantz.     Occasional.
NASTURTIUM, R. Br.     Water-Cress.
    *officinale, R. Br.   True Water-Cress.     Common.
    palustre, DC.   Marsh Cress.     "
    *   "   "   var. hispidum, DC.     Not rare.
    *Armoracia, Fries.   Horseradish.     Escaped.
BARBAREA, R. Br.     Winter Cress.
    vulgaris, R. Br.   Common Winter Cress.
SISYMBRIUM, Tourn.     Hedge Mustard.
    *officinale, Scop.   Hedge Mustard.     Worcester.
    *Prof. Eaton.*
BRASSICA, Tourn.     Mustard.
    *nigra, Koch.   Black Mustard.     Waste places.
CAPSELLA, Medic.     Shepherd's Purse.
    *Bursa-pastoris, Mœnch.     Very common.
THLASPI, Tourn.     Pennycress.
    *arvense, L.   Field Pennycress.   Waste places.   Rare.
LEPIDIUM, Tourn.     Peppergrass.
    Virginicum, L.   Wild Peppergrass.     Common.
    *ruderale, L.     Not common.
RAPHANUS, Tourn.     Radish.
    *Raphanistrum, L.   Wild Radish.   Lunenburg.
    *Mr. Kilburn.*
    *sativus, L.   Garden Radish.   Occasionally spontaneous.

## CISTACEÆ.

HELIANTHEMUM, Tourn.     Rock-rose.
    Canadense, Michx.   Frost-weed.   Common in dry grounds.

LECHEA, Kalm.   Pinweed.
major, Michx.                Templeton.   *V. P. Parkhurst.*
minor, L.                    Dry and sterile grounds.
tenuifolia, Michx.           With the last.

## VIOLACEÆ.

VIOLA, Tourn.   Violet.
pedata, L.   Bird-foot Violet.   Common in sandy soil.
palmata, L.   Common Blue Violet.   Common.
"   " var. cucullata, Gray.   Low grounds, common.
sagittata, Ait..   Arrow-leaved Violet.   Dry grounds, common.
blanda, Willd.   Sweet White Violet.   Common in wet grounds.
primulæfolia, L.   Worcester.   *Prof. Eaton.*
lanceolata, L.   Lance-leaved Violet.   Wet grounds, common.
rotundifolia, Michx.   Early Yellow Violet.   Cold woods.
pubescens, Ait.   Downy Yellow Violet.   Common.
striata, Ait.   Pale Violet.   Leicester.
rostrata, Pursh.   Long-spurred Violet.   Fitchburg.
canina, L., var. Muhlenbergii, Gray. Dog Violet. Common.
*tricolor, L.   Pansy.   Occasionally spontaneous.

## CARYOPHYLLACEÆ.

DIANTHUS, L.   Pink.
*Armeria, L. . Deptford Pink.   Dry pastures.
*deltoides, L.   Maiden Pink.   Rare.
*barbatus, L.   Sweet William.   Princeton.

SAPONARIA, L. Soapwort.
 *officinalis, L. Soapwort. Naturalized.
SILENE, L. Catchfly.
 *Cucubalus, Wibel. Bladder Campion. Meadows, not rare.
 Pennsylvanica, Michx. Wild Pink. Not abundant.
 antirrhina, L. Sleepy Catchfly. " "
 *noctiflora, L. Night-flowering Catchfly. Rare.
 *Dr. G. C. Webber.*
LYCHNIS, Tourn. Cockle.
 *vespertina, Sibth. Evening Lychnis. Occasional.
 *Githago, Lam. Corn Cockle. "
 *Flos-cuculi, L. Ragged Robin. "
ARENARIA, L. Sandwort.
 Grœnlandica, Spreng. Mountain Sandwort. Ashburnham. *Prof. Vose.*
 lateriflora, L. Sandwort. Not rare.
STELLARIA, L. Chickweed.
 *media, Smith. Common Chickweed. Very common.
 longifolia, Muhl. Long-leaved Stitchwort. Common.
 borealis, Bigel. Northern Stitchwort. In shaded places.
CERASTIUM, L. Mouse-ear Chickweed.
 *viscosum, L. Mouse-ear Chickweed. Not common.
 *vulgatum, L. Larger Mouse-ear Chickweed. Common.
 arvense, L. Field Chickweed. Dry places.
SAGINA, L. Pearlwort.
 procumbens, L. In shaded pastures. Rare.
BUDA, Adans. Sand-Spurrey.
 rubra, Dumort. Common in dry soil.

SPERGULA, L.   Spurrey.
  *arvensis, L.   Corn Spurrey.                          Not rare.

## PORTULACACEÆ.

PORTULACA, Tourn.   Purslane.
  *oleracea, L.   Common Purslane.   Common in gardens.
CLAYTONIA, Gronov.   Spring-Beauty.
  Caroliniana, Michx.                      Worcester, Holden.

## HYPERICACEÆ.

HYPERICUM, Tourn.   St. John's-wort.
  ellipticum, Hook.                   Wet places.   Common.
  *perforatum, L.   Common St. John's-wort.        Fields.
        Common.
  maculatum, Walt.            Common in damp places.
  mutilum, L.                         In low grounds.
  Canadense, L.               Common in low grounds.
  nudicaule, Walt.   Pine-weed.        In dry grounds.
ELODES, Adans.   Marsh St. John's-wort.
  campanulata, Pursh.        Common in swampy grounds.

## MALVACEÆ.

MALVA, L.   Mallow.
  *rotundifolia, L.   Common Mallow.              Common.
  *sylvestris, L.   High Mallow.                 Waysides.
  *crispa, L.   Curled Mallow.           Sparingly escaped.
  *moschata, L.   Musk Mallow.                "          "

## TILIACEÆ.

TILIA, Tourn.   Basswood.
  Americana, L.   Basswood.                        Common.

## LINACEÆ.

LINUM, Tourn.     Flax.
*usitatissimum, L.   Common Flax.      Occasionally spontaneous.

## GERANIACEÆ.

GERANIUM, Tourn.     Cranesbill.
  maculatum, L.   Wild Cranesbill.           Common.
  Robertianum, L.   Herb Robert.             Mt. Wachusett,
    *Prof. Eaton;* Worcester, *G. Coult.*
  Carolinianum, L.                           Cultivated grounds.
  *pusillum, L.                              Waste places. Rare.
ERODIUM, L'Her.     Storksbill.
  *cicutarium, L'Her.    Occasionally in meadows.
OXALIS, L.     Wood-Sorrel.
  Acetosella, L.   Common Wood-Sorrel. Northern part of the county.
  violacea, L.   Violet Wood-Sorrel.   Northern part of the county.
  corniculata, L., var. stricta, Sav.   Yellow Wood-Sorrel.
                                             Common.
IMPATIENS, L.     Balsam.
  pallida, Nutt.   Pale Touch-me-not.        Not rare.
  fulva, Nutt.   Spotted Touch-me-not.       Common.

## RUTACEÆ.

XANTHOXYLUM, L.     Prickly Ash.
  Americanum, Mill.   Northern Prickly Ash.   Millbury.

## ILICINEÆ.

ILEX, L.    Holly.
   verticillata, Gray.   Winterberry.              Common.
   lævigata, Gray.   Smooth Winterberry.          Swamps.
NEMOPANTHES, Raf.   Mountain Holly.
   fascicularis, Raf.                              Swampy woods.

## CELASTRACEÆ.

CELASTRUS, L.    Bitter-sweet.
   scandens, L.    Climbing Bitter-sweet.          Common.

## RHAMNACEÆ.

RHAMNUS, Tourn.    Buckthorn.
   *cathartica, L.    Common Buckthorn.            Naturalized.
CEANOTHUS, L.    New Jersey Tea.
   Americanus, L.    New Jersey Tea.              Common.

## VITACEÆ.

VITIS, Tourn.    Grape.
   Labrusca, L.    Northern Fox-Grape.             Common.
   cordifolia, Michx.    Frost Grape.              Not uncommon.
   riparia, Michx.                                 *G. B. Emerson.*
AMPELOPSIS, Michx.    Virginian Creeper.
   quinquefolia, Michx.                            Common.

## SAPINDACEÆ.

ACER, Tourn.    Maple.
   Pennsylvanicum, L.    Striped Maple.            Common.
   spicatum, Lam.    Mountain Maple.               Northern part of
      the county.

saccharinum, Wang.  Sugar Maple.           Common.
dasycarpum, Ehrh.  White Maple.            Northern part of
   the county.
rubrum, L.  Red Maple.                     Common.

## ANACARDIACEÆ.

RHUS, L.     Sumach.
   typhina, L.  Staghorn Sumach.           Common.
   glabra, L.  Smooth Sumach.              "
   copallina, L.  Dwarf Sumach.            "
   venenata, DC.  Poison Dogwood.          "
   Toxicodendron, L.  Poison Ivy.          "
      "       " var. radicans, L.          "

## POLYGALACEÆ.

POLYGALA, Tourn.       Milkwort.
   paucifolia, Willd.  Fringed Polygala.   Common.
   polygama, Walt.                         Dry sandy soil. Auburn.
   sanguinea, L.                           Common in moist places.
   Nuttallii, Torr. & Gray.                Dry soil.  Not common.
   verticillata, L.                        Dry soil, common.

## LEGUMINOSÆ.

BAPTISIA, Vent.      False Indigo.
   tinctoria, R. Br.  Wild Indigo.         Abundant.
GENISTA, L.    Woad-Waxen.
   *tinctoria, L.  Dyer's Green-weed.      Rare.
LUPINUS, Tourn.     Lupine.
   perennis, L.  Wild Lupine.              Quite common.

TRIFOLIUM, Tourn.   Clover.
  *arvense, L.   Rabbit-foot Clover.   Old fields, common.
  *pratense, L.   Red Clover.   Common in meadows.
  *repens, L.   White Clover.   "   "   "
  *hybridum, L.   Alsike Clover.   Worcester.
  *agrarium, L.   Yellow Clover.   "
  *procumbens, L.   Low Hop-Clover.   Millbury.
MELILOTUS, Tourn.   Melilot.
  *officinalis, Willd.   Yellow Melilot.   Not common.
  *alba, Lam.   White Melilot.   Common.
MEDICAGO, Tourn.   Medick.
  *sativa, L.   Alfalfa.   Spreading.
  *lupulina, L.   Black Medick.   Waste places.
  *maculata, Willd.   Spotted Medick.   "   "
  *denticulata.   "   "
TEPHROSIA, Pers.   Hoary Pea.
  Virginiana, Pers.   Goat's Rue.   Dry soil, not rare.
ROBINIA, L.   Locust-tree.
  Pseudacacia, L.   Common Locust.   Naturalized.
  viscosa, L.   Clammy Locust.   "
CORONILLA, L.
  *varia, L.   Rare.
DESMODIUM, Desv.   Tick-Trefoil.
  nudiflorum, DC.   Dry woods, not abundant.
  acuminatum, DC.   Rich woods.
  rotundifolium, DC.   Dry woods, not rare.
  canescens, DC.   Southbridge.   *L. E. Ammidown.*
  paniculatum, DC.   Copses, common.
  Canadense, DC.   Common on roadsides.

rigidum, DC. — Common.
ciliare, DC. — Dry hills, not uncommon.
Marilandicum, F. Boott. — Copses.

LESPEDEZA, Michx.  Bush-Clover.
   procumbens, Michx. — In dry soil.
   violacea, Pers. — In dry copses.
   Stuvei, Nutt. — " " "
   polystachya, Michx. — Dry hills.
   capitata, Michx. — With the last.

VICIA, Tourn.  Vetch.
   *sativa, L.  Common Vetch. — Waste places.
   *hirsuta, Koch. — Barre. *Miss Sara Lane.*
   Cracca, L. — Not abundant.

LATHYRUS, Tourn.  Vetchling.
   ochroleucus, Hook. — Barre. *Miss Sara Lane.*

APIOS, Boerh.  Ground-nut.
   tuberosa, Moench. — Low grounds, common.

PHASEOLUS, Tourn.  Kidney Bean.
   perennis, Walt.  Wild Bean. — Not common.

AMPHICARPÆA, Ell.  Hog Pea-nut.
   monoica, Nutt. — Common in copses.

CASSIA, Tourn.  Senna.
   Marilandica, L.  Wild Senna. — Not common.
   Chamæcrista, L.  Partridge Pea. — Southern part of the county.
   nictitans, L.  Wild Sensitive-Plant. — With the last.

## ROSACEÆ.

PRUNUS, Tourn.  Plum.  Cherry.
   pumila, L.  Dwarf Cherry.  Winchendon. *A. S. Allen.*

Pennsylvanica, L. f.  Wild Red Cherry.  Common in woods.
Virginiana, L.   Choke-Cherry.                  Copses.
serotina, Ehrh.  Wild Black Cherry.     With the last.
SPIRÆA, L.     Meadow-Sweet.
salicifolia, L.  Common Meadow-Sweet.   Pastures, etc.
tomentosa, L.   Hardhack.               Low grounds.
PHYSOCARPUS, Maxim.      Nine-bark.
opulifolius, Maxim.      Worcester.  *Miss E. J. Seaver.*
RUBUS, Tourn.   Bramble.
odoratus, L.   Purple Flowering-Raspberry.    Northern part of the county.
triflorus, Richardson.  Dwarf Raspberry.  Low grounds.
strigosus, Michx.   Red Raspberry.            Copses.
occidentalis, L.   Black Raspberry.              "
villosus, Ait.   High Blackberry.                "
Canadensis, L.   Low Blackberry.         Dry fields.
hispidus, L.   Running Swamp-Blackberry.  Low grounds.
DALIBARDA, L.    Dalibarda.
repens, L.               Northern part of the county.
GEUM, L.    Avens.
album, Gmelin.                  Borders of woods.
Virginianum, L.                  "     "     "
strictum, Ait.                  Moist meadows.
rivale, L.   Purple Avens.      Moist meadows, common.
WALDSTEINIA, Willd.
fragarioides, Tratt.   Barren Strawberry.  Winchendon.
*F. R. Hathaway.*
FRAGARIA, Tourn.    Strawberry.
Virginiana, Mill.       Fields and woodlands, common.
vesca, L.               Less common than the last.

POTENTILLA, L.   Cinque-foil.
   Norvegica, L.                                    Pastures.
   argentea, L.. Silvery Cinque-foil.               Common.
   palustris, Scop.  Marsh Five-Finger.             Rare.
   fructicosa, L.  Shrubby Cinque-foil.  Eastern part of the
      county.
   tridentata, Ait.  Three-toothed Cinque-foil.     Mt.
      Wachusett.
   Canadensis, L.  Common Cinque-foil.              Common.
AGRIMONIA, Tourn.    Agrimony.
   Eupatoria, L.  Common Agrimony.                  Copses.
POTERIUM, L.    Burnet.
   Canadense, Benth. & Hook.  Canadian Burnet.
ROSA, Tourn.   Rose.
   blanda, Ait.  Wild Rose.                   Not common.
   Carolina, L.      "       "                Borders of swamps.
   lucida, Ehrh.     "       "                    "       "       "
   humilis, Marsh.       Princeton.   *W. W. Bailey & J. F.
      Collins.*
   *rubiginosa, L.  Sweetbrier.                    Pastures.
PYRUS, L.   Pear.  Apple.
   *Malus, L.  Apple              Occasionally self-sown.
   *communis, L.  Pear.         •     "             "       "
   arbutifolia, L. f.  Choke-berry.  In huckleberry pastures.
      "         "     var. melanocarpa, Hook.   Common.
   Americana, DC.  American Mountain-Ash.           Mt.
      Wachusett.
   *aucuparia, Gært.  European Mountain-Ash.   Henshaw
      Pond, Leicester, spontaneous.

CRATÆGUS, L.   Hawthorn.
coccinea, L.   Scarlet-Thorn.   Common.
"      "  var. mollis, Torr. & Gray.   Fitchburg. Rare.
punctata, Jacq.   Worcester.
AMELANCHIER, Medic.   June-berry.
Canadensis, Torr. & Gray.   Shad-bush.   Common.

## SAXIFRAGACEÆ.

SAXIFRAGA, L.   Saxifrage.
Virginiensis, Michx.   Early Saxifrage.   Common.
Pennsylvanica, L.   Swamp Saxifrage.   "
TIARELLA, L.   False Mitre-wort.
cordifolia, L.   Abundant in northern part of the county.
MITELLA, Tourn.   Mitre-wort.
diphylla, L.   Common.
CHRYSOSPLENIUM, Tourn.   Golden Saxifrage.
Americanum, Schwein.   Common in wet places.
PARNASSIA, Tourn.   Grass of Parnassus.
Caroliniana, Michx.   In low grounds.
HYDRANGEA, Gronov.   Hydrangea.
arborescens, L.   Wild Hydrangea.   Barre. *Miss Sara Lane.* Spontaneous.
RIBES, L.   Currant. Gooseberry.
Cynosbati, L.   Wild Gooseberry.   Rocky woods.
rotundifolium, Michx.   Wild Gooseberry.   Princeton.
oxyacanthoides, L.   Wild Gooseberry.   Copses.
prostratum, L'Her.   Fetid Currant.   Mt. Wachusett.
floridum, L'Her.   Wild Black Currant.   Tatnuck.
rubrum, L., var. subglandulosum, Maxim.   Red Currant. Winchendon.

FLORA OF WORCESTER COUNTY. 273

## CRASSULACEÆ.

PENTHORUM, Gronov.   Ditch Stone-crop.
  sedoides, L.                               Common in low lands.
SEDUM, Tourn.   Stone-crop.
  ternatum, Michx.                       Worcester.   *Prof. Eaton.*
  *acre, L.   Mossy Stone-crop.                       Established.
  *Telephium, L.   Live-for-ever.                          "

## DROSERACEÆ.

DROSERA, L.   Sundew.
  rotundifolia, L.   Round-leaved Sundew.   Bogs, common.
  intermedia, Hayne, var. Americana, DC.   "      "

## HAMAMELIDEÆ.

HAMAMELIS, L.   Witch-Hazel.
  Virginiana, L.                             Copses, common.

## HALORAGEÆ.

MYRIOPHYLLUM, Vaill.   Water-Milfoil.
  spicatum, L.                                       Ponds.
  verticillatum, L.                           Ponds, common.
PROSERPINACA, L.   Mermaid-weed.
  palustris, L.                             Brooks.   Not rare.
CALLITRICHE, L.   Water-Starwort.
  verna, L.                              In brooks.   Millbury.

## MELASTOMACEÆ.

RHEXIA, L.   Meadow-Beauty.
  Virginica, L.                  Sutton.   *Dr. G. C. Webber.*

## LYTHRACEÆ.

LYTHRUM, L.   Loosestrife.
   *Salicaria, L.   Spiked Loosestrife.   Not common.
DECODON, Gmel.   Swamp Loosestrife.
   verticillatus, Ell.   Swamps.   Not common.

## ONAGRACEÆ.

LUDWIGIA, L.   False Loosestrife.
   alternifolia, L.   Seed-box.   Swamps.
   palustris, Ell.   Water Purslane.   Swamps, common.
EPILOBIUM, L.   Willow-herb.
   angustifolium, L.   Fire-weed.   Common in clearings.
   lineare, Muhl.   In low lands.
   strictum, Muhl.   "   "   "
   coloratum, Muhl.   In low lands, common.
   adenocaulon, Haussk.   Wet grounds.
ŒNOTHERA, L.   Evening Primrose.
   biennis, L.   Common Evening Primrose.   Common.
   "   "   var. grandiflora, Lindl.   Rare.
   pumila, L.   Dry fields.   Quite common.
   fruticosa, L.   Sundrops.
GAURA, L.   Gaura.
   biennis, L.   Southbridge.   *L. E. Ammidown*.
CIRCÆA, Tourn.   Enchanter's Nightshade.
   Lutetiana, L.   Common in thickets.
   alpina, L.   Sutton, Spencer.   *Miss A. E. Tucker*.

## CUCURBITACEÆ.

SICYOS, L.   One-seeded Bur-Cucumber.
   angulatus, L.   Thickets.

ECHINOCYSTIS, Torr. & Gray.    Wild Balsam-apple.
  lobata, Torr. & Gray.                              Rich soil.

## FICOIDEÆ.

MOLLUGO, L.    Indian-Chickweed.
  *verticillata, L.  Carpet-weed.                    Common.

## UMBELLIFERÆ.

DAUCUS, Tourn.    Carrot.
  *Carota, L.                                        Common.
ANGELICA, L.    Angelica.
  atropurpurea, L.                              In low grounds.
HERACLEUM, L.    Cow-Parsnip.
  lanatum, Michx.                               In low grounds.
PASTINACA, L.    Parsnip.
  *sativa, L.                                        Common.
THASPIUM, Nutt.    Meadow-Parsnip.
  aureum, Nutt.                         Common in low meadows.
CRYPTOTÆNIA, DC.    Honewort.
  Canadensis, DC.                       Worcester, Southbridge.
SIUM, Tourn.    Water Parsnip.
  cicutæfolium, Gmelin.                         In wet places.
  Carsonii, Durand.                           Swamps.  Sutton.
ZIZIA, Koch.    Zizia.
  aurea, Koch.                                 Upland meadows.
CARUM, L.    Caraway.
  *Carui, L.  Caraway.                               Common.
CICUTA, L.    Water-Hemlock.
  maculata, L.  Musquash Root.          Common in swamps.
  bulbifera, L.                    Worcester.  *H. H. Kingsbury.*

CONIUM, L.    Poison Hemlock.
   *maculatum, L.                                   Not rare.
OSMORRHIZA, Raf.    Sweet Cicely.
   brevistylis, DC.                    Worcester.  *Prof. Eaton.*
   longistylis, DC.                        "           "      "
HYDROCOTYLE, Tourn.    Water Pennywort.
   Americana, L.                       Common in damp woods.
SANICULA, Tourn.    Sanicle.
   Marylandica, L.                         Common in copses.
   "      "  var. Canadensis, Torr.        Worcester.

## ARALIACEÆ.

ARALIA, Tourn.    Ginseng.
   racemosa, L.    Spikenard.                       Occasional.
   hispida, Vent.  Wild Elder.           Common in clearings.
   nudicaulis, L.  Wild Sarsaparilla.               Common.
   quinquefolia, Decsne. & Planch.  Ginseng.    Not common.
   trifolia, Decsne. & Planch.   Dwarf Ginseng.    Common.

## CORNACEÆ.

CORNUS, Tourn.    Cornel.  Dogwood.
   Canadensis, L.   Bunch-berry.       Common northwards.
   florida, L.      Flowering Dogwood.    "    southwards.
   circinata, L'Her.  Round-leaved Cornel.   Not common.
   sericea, L.      Silky Cornel.                   Common.
   stolonifera, Michx.  Red-osier Dogwood.         Sutton.
   paniculata, L'Her.   Panicled Cornel.           Common.
   alternifolia, L. f.  Alternate-leaved Cornel.      "
NYSSA, L.    Tupelo.
   sylvatica, Marsh.    Tupelo.                   Occasional.

## Division II. GAMOPETALÆ.

### CAPRIFOLIACEÆ.

Sambucus, Tourn.   Elder.
  Canadensis, L.   Common Elder.   Common in rich soil.
  racemosa, L.   Red-berried Elder.   Rocky woods.
Viburnum, L.   Arrow-wood.
  lantanoides, Michx.   Hobble-bush.   Common in the northern part of the county.
  Opulus, L.   Cranberry-tree.   Rare in the southern part of the county.
  acerifolium, L.   Maple-leaved Viburnum.   Common.
  dentatum, L.   Arrow-wood.   Common in wet grounds.
  cassinoides, L.   Withe-rod.   Swamps, common.
  Lentago, L.   Sweet Viburnum.   Occasional.
Triosteum, L.   Fever-wort.
  perfoliatum, L.   Boylston.
Linnæa, Gronov.   Twin-flower.
  borealis, L.   Gardner, Templeton.
Symphoricarpos, Dill.   Snowberry.
  racemosus, Michx.   Snowberry.   Farnumsville. *Miss K. I. Fish.*
Lonicera, L.   Honeysuckle.
  ciliata, Muhl.   Fly-Honeysuckle.   Common.
  cærulea, L.   Mountain Fly-Honeysuckle.   Boylston.
  hirsuta, Eaton.   Hairy Honeysuckle.   *Bigelow's Fl. Bost.*
  glauca, Hill.   West Boylston, North Brookfield.
Diervilla, Tourn.   Bush-Honeysuckle.
  trifida, Moench.   Common in copses.

## RUBIACEÆ.

HOUSTONIA, L.  Houstonia.
 cærulea, L. Bluets. Common.
CEPHALANTHUS, L. Button-bush.
 occidentalis, L. Button-bush. Common in swamps.
MITCHELLA, L. Partridge-berry.
 repens, L. Partridge-berry. Common.
GALIUM, L. Bedstraw.
 *Mollugo, L. Princeton. *W. W. B. & J. F. C.*
 circæzans, Michx. Wild Liquorice. Rich woods.
 lanceolatum, Torr. Wild Liquorice. Dry woods.
  Worcester. *Miss Sargent.*
 trifidum, L. Small Bedstraw. In wet grounds.
  "   " var. pusillum, Gray. "   "   "
 asprellum, Michx. Rough Bedstraw. "   "   "
 triflorum, Michx. Sweet-scented Bedstraw. Rich woods.

## DIPSACEÆ.

DIPSACUS, Tourn. Teasel.
 *sylvestris, Mill. Wild Teasel. Worcester.

## COMPOSITÆ.

VERNONIA, Schreb. Iron-weed.
 Noveboracensis, Willd. In swampy meadows.
MIKANIA, Willd. Climbing Hemp-weed.
 scandens, L. Tatnuck. *Mr. Kinney.* Rare.
EUPATORIUM, Tourn. Thoroughwort.
 purpureum, L. Joe-Pye Weed. Common.
 rotundifolium, L., var. ovatum, Torr. Worcester. *G. Coult.*

perfoliatum, L.   Thoroughwort.   Common.
ageratoides, L.   White Snake-root. Rich woods, common.
aromaticum, L.   Southbridge.   *L. E. Ammidown.*
LIATRIS, Schreb.   Blazing-Star.
   scariosa, Willd.   Sterling.   *Miss E. J. Seaver.*
SOLIDAGO, L.   Golden-rod.
   squarrosa, Muhl.   Southbridge.   *L. E. Ammidown.*
   cæsia, L.   Rich woodlands, common.
   latifolia, L.   "   "   "
   bicolor, L.   White Golden-rod.   Common.
   puberula, Nutt.   Princeton.   *W. W. B. & J. F. C.*
   uliginosa, Nutt.   Swamps.
   speciosa, Nutt.   Barre.   *Miss Sara Lane.*
   odora, Ait.   Sweet Golden-rod.   Copses.
   patula, Muhl.   Swamps, not rare.
   rugosa, Mill.   Very common.
   ulmifolia, Muhl.   Southbridge.   *L. E. Ammidown.*
   Elliottii, Torr. & Gray.   Swamps.
   neglecta, Torr. & Gray.   "
   arguta, Ait.   Copses.
   juncea, Ait.   Copses and banks, common.
   serotina, Ait.   Copses and along fences,   "
   "   "   var. gigantea, Gray.   Near the last.
   Canadensis, L.   Very common.
   nemoralis, Ait.   Dry sterile fields,   "
   lanceolata, L.   Moist soil,   "
   tenuifolia, Pursh.   Southbridge.   *L. E. Ammidown.*
SERICOCARPUS, Nees.   White-topped Aster.
   conyzoides, Nees.   Dry grounds, common.
   solidagineus, Nees.   Worcester, Boylston.

ASTER, L.   Aster.
| | |
|---|---|
| corymbosus, Ait. | Woodlands, common. |
| macrophyllus, L. | Moist woods, " |
| Novæ-Angliæ, L. | Moist grounds, not rare. |
| patens, Ait. | Dry grounds, common. |
| undulatus, L. | Copses, " |
| cordifolius, L. | Dry woodlands, " |
| sagittifolius, Willd. | Pastures, etc., " |
| lævis, L. | Borders of woodlands, " |
| ericoides, L. | Southbridge. *L. E. Ammidown.* |
| multiflorus, Ait. | Dry sandy soil, common. |
| dumosus, L. | Copses, " |
| vimineus, Lam. | Waste grounds, " |
| diffusus, Ait. | "        "        " |
| Tradescanti, L. | Low    "        " |
| paniculatus, Lam. | Shady moist banks, " |
| salicifolius, Ait. | Pastures and low grounds, " |
| Novi-Belgii, L. | Moist grounds, " |
| prenanthoides, Muhl. | Grafton. *Misses Putnam.* |
| puniceus, L. | Low grounds, very common. |
| umbellatus, Mill. | Moist thickets, common. |
| linariifolius, L. | Dry grounds, " |
| ptarmicoides, Torr. and Gray. | Barre. *Miss Sara Lane.* |
| acuminatus, Michx. | Rich woodlands. |

ERIGERON, L.   Fleabane.
| | |
|---|---|
| Canadensis, L.   Horse-weed. | Common. |
| annuus, Pers.   Daisy Fleabane. | Waste places. |
| strigosus, Muhl.   "        " | "        " |
| "        "   var. discoideus, Robbins. | Millbury. |

FLORA OF WORCESTER COUNTY. 281

bellidifolius, Muhl.   Robin's Plantain.           Common.
Philadelphicus, L.   Common Fleabane.          "
ANTENNARIA, Gærtn.   Everlasting.
  plantaginifolia, Hook.               Pastures, common.
ANAPHALIS, DC.   Everlasting.
  margaritacea, Benth. & Hook.   Pearly Everlasting.
'                                              Common.
GNAPHALIUM, L.   Cudweed.
  polycephalum, Michx.   Common Everlasting.   Common.
  decurrens, Ives.   Everlasting.             "
  uliginosum, L.   Low Cudweed.          Low grounds.
  purpureum, L.   Purplish Cudweed.       "       "
INULA, L.   Elecampane.
  *Helenium, L.             Roadsides and low pastures.
AMBROSIA, Tourn.   Ragweed.
  artemisiæfolia, L.   Roman Wormwood.   Very common.
XANTHIUM, Tourn.   Cocklebur.
  *Strumarium, L.                      Worcester. *J. Coulson.*
HELIOPSIS, Pers.   Ox-eye.
  lævis, Pers.       Worcester.   *G. Coult.*   From the West.
RUDBECKIA, L.   Cone-flower.
  laciniata, L.           Leicester.   *H. H. Kingsbury.*
  hirta, L.            From the West.   Meadows, common.
HELIANTHUS, L.   Sunflower.
  annuus, L.   Common Sunflower.             Escaped.
  divaricatus, L.                     Thickets, not rare.
  strumosus, L.                        "      common.
  decapetalus, L.                     Copses, not common.
  tuberosus, L.   Jerusalem Artichoke.    "       "

36

BIDENS, L.   Bur-Marigold.
  frondosa, L.   Common Beggar-ticks.   Waste grounds.
  connata, Muhl.   Swamp Beggar-ticks.   Low   "
  cernua, L.   Smaller Bur-Marigold.   Wet places.
  chrysanthemoides, Michx.   Larger Bur-Marigold.   Swamps.
  Beckii, Torr.   Water Marigold.   Lake Quinsigamond.
GALINSOGA, Ruiz & Pavon.   Galinsoga.
  *parviflora, Cav.   Worcester, Millbury.
ANTHEMIS, L.   Chamomile.
  *Cotula, DC.   May-weed.   Roadsides, common.
ACHILLEA, L.   Yarrow.
  Millefolium, L.   Common Yarrow.   Common.
CHRYSANTHEMUM, Tourn.   Ox-eye Daisy.
  *Leucanthemum, L.   Ox-eye Daisy.   Very common.
  *Parthenium, Pers.   Feverfew.   Barre. *Miss Sara Lane.*
TANACETUM, L.   Tansy.
  *vulgare, L.   Common Tansy.   Quite common.
  *   "   "   var. crispum.   *G. E. Stone.*
ARTEMISIA, L.   Wormwood.
  *vulgaris, L.   Common Mugwort.   Worcester.
                                            *Prof. Eaton.*
  *Absinthium, L.   Wormwood.   Worcester.   "   "
TUSSILAGO, Tourn.   Coltsfoot.
  *Farfara, L.   Fitchburg.   *E. A. Hartwell.*
SENECIO, Tourn.   Groundsel.
  aureus, L.   Squaw-weed.   Very common.
ERECHTITES, Raf.   Fireweed.
  hieracifolia, Raf.   Common in .burned clearings.
ARCTIUM, L.   Burdock.
  *Lappa, L., var. minus.   Roadsides and waste places.

CNICUS, Tourn.     Thistle.
  *lanceolatus, Hoffm.   Common Thistle.      Common.
  horridulus, Pursh.      Southbridge.   *L. E. Ammidown.*
  altissimus, Willd., var. discolor, Gray.         Not rare.
  muticus, Pursh. Swamp Thistle.  Worcester. *Prof. Eaton.*
  pumilus, Torr.   Pasture Thistle.    Southbridge.
                                              *L. E. Ammidown.*
  *arvensis, Hoffm.   Canada Thistle.           Common.
KRIGIA, Schreber.     Dwarf Dandelion.
  Virginica, Willd.          Upland woods and pastures.
CICHORIUM, Tourn.    Chicory.
  *Intybus, L.                        Roadsides, not rare.
LEONTODON, L.    Fall Dandelion.
  *autumnalis, L.     Meadows and roadsides.   Common
    in southern part of the county.
HIERACIUM, Tourn.     Hawkweed.
  *aurantiacum, L.       Winchendon.   *A. S. Allen.*
  Canadense, Michx.                  Copses, common.
  paniculatum, L.                        "       "
  venosum, L.   Rattlesnake-weed.       In dry woods.
  scabrum, Michx.                    Dry open woods.
  Gronovii, L.   Hairy Hawkweed.           Common.
PRENANTHES, Vaill.    Rattlesnake-root.
  alba, L.   White Lettuce.      Rich woods, common.
  serpentaria, Pursh.   Lion's-foot.      *G. E. Stone.*
  altissima, L.                   Rich woods, not rare.
TARAXACUM, Haller.     Dandelion.
  *officinale, Weber.   Common Dandelion.   Everywhere.
LACTUCA, Tourn.    Lettuce.
  Canadensis, L.   Wild Lettuce.     Rich soil, common.

hirsuta, Muhl.  Southbridge. *L. E. Ammidown*
leucophæa, Gray.  Low grounds, not rare
SONCHUS, L.  Sow-Thistle.
*oleraceus, L.  Common Sow-Thistle.  Waste places
*asper, Vill.  Spiny-leaved Sow-Thistle.  " "
*arvensis, L.  Field Sow-Thistle.  " "

## LOBELIACEÆ.

LOBELIA, L.  Lobelia.
cardinalis, L.  Cardinal-flower.  Low grounds
syphilitica, L.  Great Lobelia.  " "
spicata, Lam.  Moist sandy soil
inflata, L.  Indian Tobacco.  Low grounds
Dortmanna, L.  Water Lobelia.  Borders of ponds

## CAMPANULACEÆ.

SPECULARIA, Heister.  Venus' Looking-glass.
perfoliata, A. DC.  Dry open grounds
CAMPANULA, Tourn.  Bellflower.
*rapunculoides, L.  Naturalized.
rotundifolia, L.  Harebell.  Winchendon. *A. S. Allen.*
aparinoides, Pursh.  Marsh Bellflower.  Wet meadows

## ERICACEÆ.

GAYLUSSACIA, HBK.  Huckleberry.
dumosa, Torr. & Gray.  Dwarf Huckleberry.  Northborough. *Arba Pierce.*
frondosa, Torr. & Gray.  Dangleberry. Copses, common.
resinosa, Torr. & Gray.  Black Huckleberry.  Pastures, common.

# FLORA OF WORCESTER COUNTY. 285

VACCINIUM, L. Blueberry. Cranberry.
    Pennsylvanicum, Lam. Dwarf Blueberry. Common.
    Canadense, Kalm. Winchendon. *A. S. Allen.*
    vacillans, Solander. Low Blueberry. Common.
    corymbosum, L. High Blueberry. "
    " " var. atrococcum, Gray. "
    Oxycoccus, L. Small Cranberry. Bogs, "
    macrocarpon, Ait. Large Cranberry. " "
CHIOGENES, Salisb. Creeping Snowberry.
    serpyllifolia, Salisb. Swamps. Auburn.
ARCTOSTAPHYLOS, Adans. Bearberry.
    Uva-ursi, Spreng. Bare hills, not common.
EPIGÆA, L. Trailing Arbutus.
    repens, L. Woods and pastures, common.
GAULTHERIA, Kalm. Wintergreen.
    procumbens, L. Checkerberry. Common.
ANDROMEDA, L. Andromeda.
    polifolia, L. Formerly found in Westborough; now in Whitehall Pond, just over the line into Middlesex County. It will undoubtedly be found again in this county.
    ligustrina, Muhl. Copses, common.
LEUCOTHOË, Don. · Leucothoë.
    racemosa, Gray. Millbury. Not common.
CASSANDRA, Don. Leather-Leaf.
    calyculata, Don. Swamps, common.
KALMIA, L. American Laurel.
    latifolia, L. Mountain Laurel. Common.
    angustifolia, L. Sheep Laurel. "
    glauca, Ait. Pale Laurel. Swamps, not rare.

RHODODENDRON, L.   Rose Bay. Azalea.
  viscosum, Torr.   White Swamp-Honeysuckle.   Common.
  nudiflorum, Torr.   Swamp Pink. "
  Rhodora, Don.   Rhodora.   Swamps, "
  maximum, L.   Rhododendron.   Auburn, Sturbridge.
LEDUM, L.   Labrador Tea.
  latifolium, Ait.   Worcester, Hubbardston.
CLETHRA, Gronov.   White Alder.
  alnifolia, L.   Sweet Pepperbush.   Wet copses, common.
CHIMAPHILA, Pursh.   Pipsissewa.
  umbellata, Nutt.   Prince's Pine.   Dry woods, common.
  maculata, Pursh.   Spotted Wintergreen.   With the last.
MONESES, Salisb.   One-flowered Pyrola.
  grandiflora, Salisb.   Abundant in northern part of the county. *F. L. Palmer.*
PYROLA, Tourn.   Wintergreen.
  secunda, L.   Copses, not common.
  chlorantha, Swartz.   Woods, common.
  elliptica, Nutt. Shin-leaf. "
  rotundifolia, L.   Copses, "
MONOTROPA, L.   Indian Pipe.
  uniflora, L.   Indian Pipe.   Rich woods, common.
  Hypopitys, L.   Pine-sap.   With the last.

## PRIMULACEÆ.

TRIENTALIS, L.   Chickweed-Wintergreen.
  Americana, Pursh.   Star-flower.   Woods, common.
STEIRONEMA, Raf.   Loosestrife.
  ciliatum, Raf.   Low grounds, common.
  lanceolatum, Gray.   Low grounds, not "

LYSIMACHIA, Tourn.      Loosestrife.
*vulgaris, L.                    Bolton.  *Miss J. M. Nichols.*
quadrifolia, L.                               Thickets, common.
stricta, Ait.                           Low grounds,      "
*nummularia, L.  Moneywort.                   Escaped.
thyrsiflora, L.  Tufted Loosestrife.   Northborough.
                                              *Arba Pierce.*
ANAGALLIS, Tourn.      Pimpernel.
*arvensis, L.                   Waste grounds.  Southbridge.

## OLEACEÆ.

FRAXINUS, Tourn.     Ash.
Americana, L.   White Ash.                    Common.
sambucifolia, Lam.   Black Ash.               Not rare.

## APOCYNACEÆ.

APOCYNUM, Tourn.     Dogbane.
androsæmifolium, L.  Spreading Dogbane.   Not rare.
cannabinum, L.   Indian Hemp.                   "    "

## ASCLEPIADACEÆ.

ASCLEPIAS, L.    Milkweed.
tuberosa, L.   Butterfly-weed.                Millbury.
purpurascens, L.   Purple Milkweed.           Roadsides.
incarnata, L.   Swamp Milkweed.               Swamps.
   "       "  var. pulchra, Pers.             With the last.
Cornuti, Decsne.   Common Milkweed.           Common.
obtusifolia, Michx.                           Not common.
phytolaccoides, Pursh.  Poke Milkweed.   Moist copses.
quadrifolia, L.                         Millbury, Worcester.

ACERATES, Ell. Green Milkweed.
viridiflora, Ell. In the western part of the county.

## GENTIANACEÆ.

GENTIANA, Tourn. Gentian.
crinita, Frœl. Fringed Gentian. Common.
Andrewsii, Griseb. Closed Gentian. "
MENYANTHES, Tourn. Buckbean.
trifoliata, L. Sutton, Paxton, etc.
LIMNANTHEMUM, Gmelin. Floating Heart.
lacunosum, Griseb. Lake Quinsigamond. Crystal Lake, Gardner. *F. L. Palmer.*

## POLEMONIACEÆ.

POLEMONIUM, Tourn. Greek Valerian.
reptans, L. Rare. *Miss G. Hakes.*

## BORRAGINACEÆ.

CYNOGLOSSUM, Tourn. Hound's-Tongue.
*officinale, L. Common Hound's-Tongue. Waste grounds.
Virginicum, L. Wild Comfrey. Millbury, Princeton.
ECHINOSPERMUM, Lehm. Stickseed.
Virginicum, Lehm. Beggar's Lice. Common.
MYOSOTIS, Dill. Forget-me-not.
laxa, Lehm. Spencer, Lunenburg.
LITHOSPERMUM, Tourn. Gromwell.
*arvense, L. Corn Gromwell. Occasional.
*officinale, L. Common Gromwell. *Miss E. J. Seaver.*

SYMPHYTUM, Tourn.    Comfrey.
*officinale, L.   Common Comfrey.             Escaped.
ECHIUM, Tourn.    Viper's Bugloss.
*vulgare, L.  Blue-weed.  Southbridge. *L. E. Ammidown.*

## CONVOLVULACEÆ.

CONVOLVULUS, Tourn.    Bindweed.
   spithamæus, L.                            Not rare.
   sepium, L.  Hedge Bindweed.                 "     "
   *arvensis, L.                             Old fields.
CUSCUTA, Tourn.    Dodder.
   Gronovii, Willd.               Wet places.  Common.

## SOLANACEÆ.

SOLANUM, Tourn.    Nightshade.
   *Dulcamara, L.   Bittersweet.              Common.
   nigrum, L.   Common Nightshade.          Southbridge.
PHYSALIS, L.    Ground Cherry.
   Virginiana, Mill.           Roadsides.  Not common.
NICANDRA, Adans.    Apple of Peru.
   *physaloides, Gærtn.    Waste grounds.  *G. E. Stone.*
DATURA, L.    Thorn-Apple.
   *Stramonium, L.         Waste grounds.  Not rare.
   *Tatula, L.                "        "      "     "

## SCROPHULARIACEÆ.

VERBASCUM, L.    Mullein.
   *Thapsus, L.  Common Mullein.      Fields.  Common.
   *Blattaria.   Moth Mullein.    Roadsides.  Not common.

37

LINARIA, Tourn.     Toad-Flax.
   Canadensis, Dumont.     ·     Sandy soil.     Common.
   *vulgaris, Mill.     Butter and Eggs.     "
CHELONE, Tourn.     Snake-head.
   glabra, L.     Wet places.     Common.
PENSTEMON, Mitchell.     Beard-tongue.
   pubescens, Solander.     Worcester.     *Miss E. F. Brown.*
MIMULUS, L.     Monkey-flower.
   ringens, L.     Wet places.     Common.
GRATIOLA, L.     Hedge-Hyssop.
   Virginiana, L.     Lake Quinsigamond.     *Prof. Eaton.*
   aurea, Muhl.     Sandy swamps.     Common.
ILYSANTHES, Raf.     False Pimpernel.
   riparia, Raf.     Wet places.
VERONICA, L.     Speedwell.
   Anagallis, L.     Water Speedwell.
   scutellata, L.     Marsh Speedwell.     Bogs.     Common.
   officinalis, L.     Common Speedwell.     Dry hills.     "
   *Chamædrys, L.     *G. E. Stone.*
   serpyllifolia, L.     Thyme-leaved Speedwell.     Common.
   *arvensis, L.     Corn Speedwell.     Rather rare.
   *agrestis, L.     Field Speedwell.     Southbridge.
                                                *L. E. Ammidown.*
GERARDIA, L.     False Foxglove.
   pedicularia, L.     Woods and thickets.
   flava, L.     Downy False Foxglove.     Common in dry woods.
   quercifolia, Pursh.     Smooth False Foxglove.     With the first.

purpurea, L.   Purple Gerardia.         In low grounds.
tenuifolia, Vahl.   Slender Gerardia.   Southbridge.
CASTILLEIA, Mutis.   Painted-Cup.
   coccinea, Spreng.   Scarlet Painted-Cup.   Occasional.
SCHWALBEA, Gronov.   Chaff-seed.
   Americana, L.        Southbridge.   *L. E. Ammidown.*
PEDICULARIS, Tourn.   Lousewort.
   Canadensis, L.   Common Lousewort.   Common.
   lanceolata, Michx.                   Rather rare.
MELAMPYRUM, Tourn.   Cow-Wheat.
   Americanum, Michx.        Open woods.   Common.

## OROBANCHACEÆ.

EPIPHEGUS, Nutt.   Beech-drops.
   Virginiana, Bart.        Ashburnham.   *Prof. Vose.*
CONOPHOLIS, Wallroth.   Squaw-root.
   Americana, Wallroth.      Ashburnham.   *Prof. Vose.*
APHYLLON, Mitchell.   Naked Broom-rape.
   uniflorum, Gray.   One-flowered Cancer-root.   Common.

## LENTIBULARIACEÆ.

UTRICULARIA, L.   Bladderwort.
   inflata, Walt.                      Ponds.  Not rare.
   vulgaris, L.   Greater Bladderwort.   Ponds, common.
   minor, L.   Smaller      "        Ponds.   *G. E. Stone.*
   gibba, L.                           Ponds.   *Miss E. J. Seaver.*
   intermedia, Hayne.                  Pools.   *G. E. Stone.*
   purpurea, Walt.   Lake Quinsigamond.   "   "   "
   cornuta, Michx.             Lunenburg.   *F. L. Palmer.*

## PEDALIACEÆ.

MARTYNIA, L. Unicorn-plant.
    proboscidea, Glox.      Worcester. *Prof. Eaton.*

## VERBENACEÆ.

VERBENA, Tourn.    Vervain.
    urticæfolia, L.   White Vervain.      Waste grounds.
    hastata, L.   Blue Vervain.               "    "
PHRYMA, L.    Lopseed.
    Leptostachya, L.      Open woods. Not abundant.

## LABIATÆ.

TRICHOSTEMA, L.    Blue Curls.
    dichotomum, L.   Bastard Pennyroyal.    Sandy fields.
COLLINSONIA, L.    Horse-Balm.
    Canadensis, L.   Stone-root.      Rich woods.
MENTHA, Tourn.    Mint.
    *viridis, L.   Spearmint.      Not rare.
    *piperita, L.   Peppermint.      Common.
    *aquatica, L.   Water Mint.   Winchendon. *A. S. Allen.*
    *arvensis, L.   Corn Mint.   Southbridge. *L. E. Ammi-down.*
    Canadensis, L.   Wild Mint.
LYCOPUS, Tourn.    Water Horehound.
    Virginicus, L.   Bugle-weed.      In low grounds.
    sessilifolius, Gray.      Not common.
    sinuatus, Ell.    In low grounds. Quite common.
PYCNANTHEMUM, Michx.    Mountain Mint.
    lanceolatum, Pursh.      Spencer. *J. C. Lyford.*

linifolium, Pursh. *G. E. Stone.*
muticum, Pers. Mt. Wachusett. *Mrs. G. B. Stearns.*
THYMUS, Tourn. Thyme.
　*Serpyllum, L. Creeping Thyme. Worcester.
CALAMINTHA, Tourn. Calamint.
　Clinopodium, Benth. Basil. Millbury.
MELISSA, L. Balm.
　*officinalis, L. Common Balm. Escaped. *G. E. Stone.*
HEDEOMA, Pers. Mock Pennyroyal.
　pulegioides, Pers. American Pennyroyal. Not rare.
NEPETA, L. Cat-Mint.
　*Cataria, L. Catnip. Near dwellings, common.
　*Glechoma, Benth. Gill-over-the-ground. Quite common.
SCUTELLARIA, L. Skullcap.
　lateriflora, L. Mad-dog Skullcap. Wet places, common.
　galericulata, L. Wet places. With the last.
BRUNELLA, Tourn. Self-heal.
　vulgaris, L. Common Self-heal. Common.
LEONURUS, L. Motherwort.
　*Cardiaca, L. Waste places, common.
LAMIUM, L. Dead-Nettle.
　*amplexicaule, L. Waste grounds. *Miss A. H. Tucker.*
　*purpureum, L. Not common.
　*maculatum, L. Escaped. *R. C. Manning.*
GALEOPSIS, L. Hemp-Nettle.
　*Tetrahit, L. Waste places.
STACHYS, Tourn. Hedge-Nettle.
　palustris, L. Wet grounds. Tatnuck. *Mr. Kinney.*

## PLANTAGINACEÆ.

PLANTAGO, Tourn. Plantain.
major, L. Common Plantain. Very common.
Rugelii, Decaisne. Princeton. *W. W. B. & J. F. C.*
*lanceolata, L. Ribgrass. Very common.
Patagonica, Jacq., var. aristata, Gray. Southbridge.
*L. E. Ammidown.*

## DIVISION III. APETALÆ.

### ILLECEBRACEÆ.

ANYCHIA, Michx. Forked Chickweed.
dichotoma, Michx. Open grounds.
SCLERANTHUS, L. Knawel.
*annuus, L. Worcester. *J. Coulson.*

### AMARANTACEÆ.

AMARANTUS, Tourn. Amaranth.
*hypochondriacus, L. Fitchburg. *Prof. Eaton.*
*retroflexus, L. Cultivated grounds, common.
albus, L. Tumble Weed. Waste grounds.

### CHENOPODIACEÆ.

CHENOPODIUM, Tourn. Pigweed.
*album, L. Very common.
hybridum, L. Maple-leaved Goosefoot. Waste places.
*Botrys, L. Jerusalem Oak. Lunenburg. *Mr. Kilburn.*

### PHYTOLACCACEÆ.

PHYTOLACCA, Tourn. Pokeweed.
decandra, L. Common Poke. Copses.

## POLYGONACEÆ.

RUMEX, L.   Dock.
*crispus, L.   Curled Dock.   Common.
*obtusifolius, L.   Bitter Dock.   Worcester.   *Prof. Eaton.*
*conglomeratus, Murray.   Smaller Green Dock.   Moist places.
*Acetosella, L.   Field Sorrel.   Common.
POLYGONUM, Tourn.   Knotweed.
aviculare, L.   Common in yards and waste places.
erectum, L.   Waysides, common.
tenue, Michx.   Worcester.   *Prof. Eaton.*
Pennsylvanicum, L.   Royalston.   "    "
amphibium, L.   North Worcester.
Muhlenbergii, Watson.   "    "
*orientale, L.   Prince's Feather.   Escaped.
*Persicaria, L.   Lady's Thumb.   *G. E. Stone.*
Hydropiper, L.   Common Smartweed.   Common.
acre, HBK.   Water Smartweed.   Wet places.
Virginianum, L.   Thickets.
arifolium, L.   Halberd-leaved Tear-thumb.   Low grounds.
sagittatum, L.   Arrow-leaved Tear-thumb.   "    "
*Convolvulus, L.   Black Bindweed.   Worcester.
*Prof. Eaton.*
cilinode, Michx.   Low grounds.   Quite common.
dumetorum, L., var. scandens, Gray.   Gardner.
*Prof. Eaton.*
FAGOPYRUM, Tourn.   Buckwheat.
*esculentum, Mœnch.   Spontaneous.   Occasional.
POLYGONELLA, Michx.   Joint-weed.
articulata, Meisn.   Mendon, Gardner.

## PODOSTEMACEÆ.

PODOSTEMON, Michx.   River-weed.
   ceratophyllus, Michx.   Shallow streams.

## ARISTOLOCHIACEÆ.

ASARUM, Tourn.   Wild Ginger.
   Canadense, L.   Millbury. Not common.

## LAURACEÆ.

SASSAFRAS, Nees.
   officinale, Nees.   Rich woods. Common.
LINDERA, Thunb.   Wild Allspice.
   Benzoin, Blume. Spice-bush.   Common.

## THYMELÆACEÆ.

DIRCA, L.   Leatherwood.
   palustris, L.   Damp woods, common.
DAPHNE, Linn.   Mezereum.
   *Mezereum, L.   Southbridge. *L. E. Ammidown.*

## SANTALACEÆ.

COMANDRA, Nutt.   Bastard Toad-flax.
   umbellata, Nutt.   Dry grounds. Common.

## EUPHORBIACEÆ.

EUPHORBIA, L.   Spurge.
   maculata, L.   Roadsides, common.
   Preslii, Guss.   Millbury.
   corollata, L.   Southbridge. *L. E. Ammidown.*

*Esula, L.                        Rare. *Miss K. I. Fish.*
*Cyparissias, L.              Escaped. Common.
ACALYPHA, L.    Three-seeded Mercury.
   Virginica, L.                Fields and open places.

## URTICACEÆ.

ULMUS, L.    Elm.
   fulva, Michx.   Slippery Elm.           Rare.
   Americana, L.   American Elm.         Common.
CELTIS, Tourn.    Nettle-tree.
   occidentalis, L.   Hackberry.             Rare.
CANNABIS, Tourn.    Hemp.
   *sativa, L.                      East Worcester.
HUMULUS, L.    Hop.
   Lupulus, L.   Common Hop.         Low grounds.
MORUS, Tourn.    Mulberry.
   *alba, L.   White Mulberry.          Spontaneous.
URTICA, Tourn.    Nettle.
   gracilis, Ait.              Waste places, common.
   *dioica, L.              Waste places. Rather rare.
LAPORTEA, Gaudichaud.    Wood-Nettle.
   Canadensis, Gaudichaud.    Rich soil. Rather rare.
PILEA, Lindl.    Richweed.
   pumila, Gray.             Moist places, common.
BŒHMERIA, Jacq.    False Nettle.
   cylindrica, Willd.           Rich soil. Quite common.

## PLATANACEÆ.

PLATANUS, L.    Buttonwood.
   occidentalis, L.                    Common.

## JUGLANDACEÆ.

JUGLANS, L.  Walnut.
  cinerea, L.  Butternut.           Quite common.
CARYA, Nutt.  Hickory.
  alba, Nutt.  Shag-bark Hickory.       Common.
  porcina, Nutt.  Pig-nut Hickory.       "
  amara, Nutt.  Bitter-nut Hickory.    Not  "

## MYRICACEÆ.

MYRICA, L.  Bayberry.
  Gale, L.  Sweet Gale.          Wet borders of ponds.
  cerifera, L.  Bayberry.          Dry sandy soil.
  asplenifolia, Endl.  Sweet Fern.    Pastures, common.

## CUPULIFERÆ.

BETULA, Tourn.  Birch.
  lenta, L.  Black Birch.           Common.
  lutea, Michx. f.  Yellow Birch.    "
  populifolia, Ait.  Gray  "          "
  papyrifera, Marshall.  Paper Birch.  Rare in southern part of the county.
  nigra, L.  Red Birch.  Fitchburg.  *E. A. Hartwell.*
ALNUS, Tourn.  Alder.
  incana, Willd.  Speckled Alder.    Common.
  serrulata, Willd.  Smooth  "       "
CORYLUS, Tourn.  Hazel-nut.
  Americana, Walt.  Wild Hazel-nut.    Common.
  rostrata, Ait.  Beaked Hazel-nut.     "
OSTRYA, Micheli.  Hop-Hornbeam.
  Virginica, Willd.                 Not rare.

CARPINUS, L.     Hornbeam.
  Caroliniana, Walter.                           Common.
QUERCUS, L.      Oak.
  alba, L.  White Oak.                           Common.
  macrocarpa, Michx.  Bur Oak.                   Worcester.
  bicolor, Willd.  Swamp White Oak.              Not rare.
  Prinus, L.  Chestnut-Oak.                      Common.
  "   "  var. monticola, Michx.                    "
  prinoides, Willd.                              Not rare.
  rubra, L.  Red Oak.                            Common.
  coccinea, Wang.  Scarlet Oak.                    "
  "   "  var. tinctoria, Gray.  Black Oak.         "
  ilicifolia, Wang.  Scrub-Oak.                    "
CASTANEA, Tourn.     Chestnut.
  sativa, Mill., var. Americana, Michx.          Common.
FAGUS, Tourn.     Beech.
  ferruginea, Ait.                               Common.

## SALICACEÆ.

SALIX, Tourn.     Willow.
  nigra, Marsh.  Black Willow.                   Not rare.
  lucida, Muhl.  Shining Willow.                 Common.
  *fragilis, L.  Crack Willow.    Worcester.  *Prof. Eaton.*
  *alba, L.  White Willow.                       Not common.
  *  "   "  var. vitellina, Koch.    Southbridge.
                                    *L. E. Ammidown.*
  rostrata, Richardson.  Princeton.  *W. W. B. & J. F. C.*
  discolor, Muhl.  Glaucous Willow.              Common.
  humilis, Marsh.  Prairie Willow.     Not       "
  tristis, Ait.  Dwarf Gray Willow.    Uxbridge.
                                        *Thos. Morong.*

sericea, Marsh.   Silky Willow.           Common.
cordata, Muhl.    Heart-leaved Willow.    Worcester.
                                          *Prof. Eaton.*
myrtilloides, L.                          Leicester.
POPULUS, Tourn.   Poplar.
tremuloides, Michx.   American Aspen.     Common.
grandidentata, Michx.  Large-toothed Aspen.   "
balsamifera, L., var. candicans, Gray.   Balm of Gilead.
    Spontaneous.
monilifera, Ait.   Cotton-wood.           Rare.

### CERATOPHYLLACEÆ.

CERATOPHYLLUM, L.   Hornwort.
  demersum, L.             Slow streams.   *G. E. Stone.*
  "       " var. echinatum, Gray.          "   "   "

### SUBCLASS II. GYMNOSPERMÆ.

### CONIFERÆ.

PINUS, Tourn.   Pine.
  Strobus, L.   White Pine.                Common.
  rigida, Mill.   Pitch  "                    "
  resinosa, Ait.   Red  "   Templeton.   *V. P. Parkhurst.*
PICEA, Link.   Spruce.
  nigra, Link.   Black Spruce.             Not rare.
TSUGA, Carrière.   Hemlock.
  Canadensis, Carr.                        Quite common.
ABIES, Link.   Fir.
  balsamea, Miller.   Balsam Fir.   Northern part of the
    county.

LARIX, Tourn.     Larch.
  Americana, Michx.   Hackmatack.           Not rare.
CHAMÆCYPARIS, Spach.   White Cedar.
  sphæroidea, Spach.                    Swamps, common.
JUNIPERUS, L.   Juniper.
  communis, L.   Common Juniper.            Not rare.
  Sabina, L., var. procumbens, Pursh.         "    "
  Virginiana, L.   Red Cedar.               "    "
TAXUS, Tourn.   Yew.
  Canadensis, Willd.                 Worcester. *G. Coult.*

## CLASS II. MONOCOTYLEDONEÆ.

### HYDROCHARIDACEÆ.

ELODEA, Michx.   Water-weed.
  Canadensis, Michx.                 Ponds, common.
VALLISNERIA, L.   Eel-grass.
  spiralis, L.                       Ponds, common.

### ORCHIDACEÆ.

MICROSTYLIS, Nutt.   Adder's Mouth.
  monophyllos, Lindl.       Spencer. *H. H. Kingsbury.*
  ophioglossoides, Nutt.                     Auburn.
LIPARIS, Richard.   Twayblade.
  liliifolia, Richard.               Millbury. Rare.
  Lœsèlii, Richard.         Spencer. *Miss A. E. Tucker.*
CORALLORHIZA, Haller.   Coral-root.
  innata, R. Br.                    Swamps. Millbury.

odontorhiza, Nutt. Southbridge. *L. E. Ammidown.*
multiflora, Nutt. Dry woods. Not rare.
SPIRANTHES, Richard. Ladies' Tresses.
latifolia, Torr. Southbridge. *L. E. Ammidown.*
cernua, Richard. Wet places, common.
præcox, Watson. Wet grassy places. Rare.
gracilis, Bigelow. Pastures. Quite common.
simplex, Gray. Winchendon. *A. S. Allen.*
GOODYERA, R. Br. Rattlesnake-Plantain.
repens, R. Br. Woods, mostly under evergreens.
pubescens, R. Br. " " " "
ARETHUSA, Gronov. Arethusa.
bulbosa, L. Bogs. Not rare.
CALOPOGON, R. Br. Grass Pink.
pulchellus, R. Br. Bogs. Not rare.
POGONIA, Juss. Pogonia.
ophioglossoides, Nutt. Common in southern part of the county.
verticillata, Nutt. Woods. Rather rare.
ORCHIS, L. Orchis.
spectabilis, L. Rich woods. Rare.
HABENARIA, Willd. Rein-Orchis.
tridentata, Hook. Worcester. *G. Coult.*
virescens, Spreng. Wet meadows. Millbury.
bracteata, R. Br. Mt. Wachusett.
obtusata, Richardson. Mt. Wachusett. *M. Pratt.*
Hookeri, Torr. Damp woods. Not rare.
orbiculata, Torr. Rich " " "
ciliaris, R. Br. Northborough, *Dr. Bigelow;* Uxbridge, *Miss Goldthwaite.*

blephariglottis, Torr.  Westminster. *Miss E. J. Seaver.*
leucophæa, Gray.  Bolton. *Miss C. Knapp.*
  Only one specimen was found, July, 1894.
lacera, R. Br.  Ragged Fringed-Orchis.  Quite common.
psycodes, Gray.  Purple Fringed-Orchis.  "      "
fimbriata, R. Br.   "        "       "       "    "

CYPRIPEDIUM, L.  Lady's Slipper.
parviflorum, Salisb.  Smaller Yellow Lady's Slipper. Rare.
pubescens, Willd.  Larger  "     "      "       "
spectabile, Swartz.  Showy Lady's Slipper. Ashburnham.
  *Prof. Vose.*
acaule, Ait.  Stemless Lady's Slipper.  Common.

## HÆMODORACEÆ.

ALETRIS, L.  Colic-root.
farinosa, L.  Grassy places. Not common.

## IRIDACEÆ.

IRIS, Tourn.  Flower-de-Luce.
versicolor, L.  Larger Blue Flag.  Common.
SISYRINCHIUM, L.  Blue-eyed Grass.
angustifolium, Mill.  Moist meadows, common.
anceps, Cav.  Spencer. *H. H. Kingsbury.*

## AMARYLLIDACEÆ.

HYPOXIS, L.  Star-grass.
erecta, L.  Meadows, common.

## LILIACEÆ.

SMILAX, Tourn.  Greenbrier.
herbacea, L.  Carrion-Flower.  Quite common.

rotundifolia, L.   Common Greenbrier.            Common.
ALLIUM, L.   Onion.
   tricoccum, Ait.   Wild Leek.                  Tatnuck.
   Canadense, Kalm.   Wild Garlic.               Not rare.
ORNITHOGALUM, Tourn.   Star-of-Bethlehem.
   *umbellatum, L.                               Escaped.
HEMEROCALLIS, L.   Day-Lily.
   *fulva, L.            Princeton. *W. W. B. & J. F. C.*
POLYGONATUM, Tourn.   Solomon's Seal.
   biflorum, Ell.                                Woods, common.
ASPARAGUS, Tourn.   Asparagus.
   *officinalis, L.                              Escaped.
SMILACINA, Desf.   False Solomon's Seal.
   racemosa, Desf.   False Spikenard.            Quite common.
   trifolia, Desf. Bogs. Rare in southern part of the county.
MAIANTHEMUM, Wigg.   False Solomon's Seal.
   Canadense, Desf.                              Moist woods, common.
STREPTOPUS, Michx.   Twisted-Stalk.
   amplexifolius, DC.     Mt. Wachusett. *M. Pratt.*
   roseus, Michx.                                Woods, common.
CLINTONIA, Raf.   Clintonia.
   borealis, Raf.                Cold moist woods, common.
UVULARIA, L.   Bellwort.
   perfoliata, L.                     Rich woods, common.
OAKESIA, Watson.   Wild Oats.
   sessilifolia, Watson.              Low woods, common.
ERYTHRONIUM, L.   Dog's-tooth Violet.
   Americanum, Ker.     Rich moist grounds, common.
LILIUM, L.   Lily.
   Philadelphicum, L.   Wood Lily.               Not rare.

superbum, L.   Turk's-cap Lily.   Southbridge.
*L. E. Ammidown.*
Canadense, L.   Wild Yellow Lily.   Moist meadows. Common.
*tigrinum, Ker.   Tiger Lily.   Escaped.
MEDEOLA, Gronov.   Indian Cucumber-root.
  Virginiana, L.   Rich woods, common.
TRILLIUM, L.   Wake Robin.
  erectum, L.   Purple Trillium.   Rich woods, common.
  grandiflorum, Salisb.   Barre.   *Miss Sara Lane.*   Rare.
  cernuum, L.   Wake Robin.   Moist woods, common.
  erythrocarpum, Michx.   Painted Trillium.   Common.
VERATRUM, Tourn.   False Hellebore.
  viride, Ait.   Indian Poke.   Common.

## PONTEDERIACEÆ.

PONTEDERIA, L.   Pickerel-weed.
  cordata, L.   Shallow water, common.

## XYRIDACEÆ.

XYRIS, Gronov.   Yellow-eyed Grass.
  flexuosa, Muhl.   Bogs, not rare.
  "   "   var. pusilla, Gray.   "   "

## JUNCACEÆ.

JUNCUS, Tourn.   Rush.
  effusus, L.   Common Rush.   Very common.
  marginatus, Rostk.   Moist sandy places.
  "   "   var. *paucicapitatus, Engelm.
            Princeton.   *W. W. B. & J. F. C.*

tenuis, Willd. Common.
pelocarpus, E. Meyer. Westminster. *Miss M. B. White.*
militaris, Bigel. Bogs and streams. Uxbridge.
acuminatus, Michx. Common.
Canadensis, J. Gay. Common.
" " " var. coarctatus, Engelm.
Princeton. *W. W. B. & J. F. C.*
LUZULA, DC. Wood-Rush.
vernalis, DC. Mt. Wachusett.
spadicea, DC., var. melanocarpa, Meyer. " "
campestris, DC. Woods, common.

## TYPHACEÆ.

TYPHA, Tourn. Cat-tail Flag.
latifolia, L. Common Cat-tail. Marshes, common.
SPARGANIUM, Tourn. Bur-reed.
eurycarpum, Engelm. Borders of ponds, common.
simplex, Huds. Southbridge. *L. E. Ammidown.*
" " var. androcladum, Engelm. *G. E. Stone.*
" " " angustifolium, " " " "

## ARACEÆ.

ARISÆMA, Martius. Indian Turnip.
triphyllum, Torr. Jack-in-the-Pulpit. Common.
PELTANDRA, Raf. Arrow Arum.
undulata, Raf. Shallow water. Sutton.
CALLA, L. Water Arum.
palustris, L. Cold bogs, common.
SYMPLOCARPUS, Salisb. Skunk Cabbage.
fœtidus, Salisb. Moist grounds, common.

ACORUS, L.    Sweet Flag.
   Calamus, L.                    Moist grounds, not rare.

## LEMNACEÆ.

LEMNA, L.    Duckweed.
   minor, L.                      Stagnant waters, common.

## ALISMACEÆ.

ALISMA, L.    Water-Plantain.
   Plantago, L.                   Shallow water, common.
SAGITTARIA, L.    Arrow-head.
   variabilis, Engelm.    In water or wet places. Common.
   "           "       var. *obtusa.           G. E. Stone.
   "           "        "  *angustifolia.       "   "    "
   "           "        "  *diversifolia.      Princeton.
                                           W. W. B. & J. F. C.
   "           "        "  gracilis, Engelm.   G. E. Stone.
   graminea, Michx.                             "   "    "

## NAIADACEÆ.

POTAMOGETON, Tourn.    Pondweed. (By the late Rev. Thos.
   Morong.)
   natans, L.                      Very common.
   "    "  var. *prolixus, Koch.     "      "
   Oakesianus, Robbins.            Uxbridge.
   Pennsylvanicus, Cham.   Common in streams and ponds.
   Vaseyi, Robbins.                Lake Quinsigamond.
   Spirillus, Tuckerm.             Common.
   hybridus, Michx.                    "
   fluitans, Roth.    Worcester.   Miss E. W. Sargent.

amplifolius, Tuckerm.        Lake Quinsigamond.
heterophyllus, Schreb.        "  "
"    "   var. myriophyllus, Robbins.
                         Lake Quinsigamond.
obtusifolius, Mertens & Koch. Beaver Brook, Worcester.
pauciflorus, Pursh.          Lake Quinsigamond.
pusillus, L.                  Common.
"   " var. tenuissimus, Koch.     "
gemmiparus, Robbins.           "
Tuckermani, Robbins.    Shockalog Pond, Uxbridge.
Robbinsii, Oakes.           Lake Quinsigamond.
NAIAS, L.    Naiad.
flexilis, Rostk & Schmidt.    Lake Quinsigamond.
                               *T. Morong.*
Indica, Cham., var. gracillima, A. Br. Lake Quinsigamond.
                               *T. Morong.*

## ERIOCAULEÆ.

ERIOCAULON, L.    Pipewort.
septangulare, With.      Borders of ponds, not rare.

## CYPERACEÆ.

CYPERUS, Tourn.    Galingale.
diandrus, Torr.                      Low grounds.
filiculmis, Vahl.                      Dry soil.
dentatus, Torr.       Margins of ponds, not rare.
strigosus, L.                            Swamps.
DULICHIUM, Pers.    Dulichium.
spathaceum, Pers.                     Swamps.

ELEOCHARIS, R. Br.     Spike-Rush.
   ovata, R. Br.                                 Swamps.
   palustris, R. Br.                           Ponds.
   intermedia, Schultes.
   tenuis, Schultes.                 Margins of ponds.
   acicularis, R. Br.            Worcester. *Prof. Eaton.*
   pygmæa, Torr.                       *G. E. Stone.*
FIMBRISTYLIS, Vahl.     Fimbristylis.
   autumnalis, Roem. & Schultes.      Low grounds.
   capillaris, Gray.                     Dry soil.
SCIRPUS, Tourn.     Bulrush.
   subterminalis, Torr.              Slow streams.
   lacustris, L.     Great Bulrush.    Ponds and streams.
   sylvaticus, L.                  Brooks. *Prof. Eaton.*
   atrovirens, Muhl.             Wet places, common.
   polyphyllus, Vahl.           Worcester. *Prof. Eaton.*
ERIOPHORUM, L.     Cotton-Grass.
   cyperinum, L.                  Swamps, common.
   alpinum, L.                     Eastern part of the county.
   vaginatum, L.                  Worcester. *G. Coult.*
   Virginicum, L.                 Low meadows, common.
   polystachyon, L.               "      "      "
   "     " var. *latifolium, Gray.   Bogs, not rare.
   gracile, Koch.                 "      "      "
RHYNCHOSPORA, Vahl.     Beak-Rush.
   glomerata, Vahl.              Low grounds, not rare.
CAREX, Ruppius.     Sedge.
   folliculata, L.                  Quite common.
   intumescens, Rudge.            Meadows.

lupulina, Muhl.     Swamps.
monile, Tuckerm.     Meadows.
bullata, Schkuhr.     "
lurida, Wahl.     "
Pseudo-Cyperus, L.     Fitchburg. *Prof. Eaton.*
"   "   " var. Americana, Hochst.     Millbury.
vestita, Willd.     Meadows.
filiformis, L.     Worcester. *Prof. Eaton.*
"   " var. latifolia, Bœckl.    "   "   "
trichocarpa, Muhl.     "   "   "
riparia, W. Curtis.     Swamps.
fusca, All.     Bogs. Southbridge.
stricta, Lam.     "
prasina, Wahl.     Meadows and bogs. Not common.
crinita, Lam.     Bogs, common.
virescens, Muhl.     Copses.
"   " var. *costata, Dewey. Princeton.
                         *W. W. B. & J. F. C.*
triceps, Michx., var. hirsuta, Bailey.    Woods.
longirostris, Torr. Princeton.    *W. W. B. & J. F. C.*
arctata, Boott.     Woods and copses.
debilis, Michx., var. Rudgei, Bailey.    Swamps.
gracillima, Schwein.     Woodlands, common.
grisea, Wahl.     Moist grounds.
flava, L.     Worcester. *Prof. Eaton.*
pallescens, L.     Glades and meadows, common.
conoidea, Schkuhr.     Grassy places.
laxiflora, Lam.     Grassy places, common.
"   " var. *latifolia, Boott. Princeton.
                         *W. W. B. & J. F. C.*

digitalis, Willd. Dry woods. *G. E. Stone.*
platyphylla, Carey. Southbridge. *L. E. Ammidown.*
plantaginea, Lam. " " " "
tetanica, Schkuhr. Wet meadows.
varia, Muhl. Dry woods.
Pennsylvanica, Lam. Dry fields, common.
umbellata, Schkuhr. Grassy knolls. Not common.
pubescens, Muhl. Copses. *G. E. Stone.*
Willdenovii, Schkuhr. Worcester. *Prof. Eaton.*
polytrichoides, Muhl. Swamps.
stipata, Muhl. Roadsides. Not rare.
teretiuscula, Gooden. Worcester. *Prof. Eaton.*
vulpinoidea, Michx. Meadows.
rosea, Schkuhr. Rich woods.
" " var. *radiata, Dewey. Princeton.
*W. W. B. & J. F. C.*
sparganioides, Muhl. Meadows.
*muricata, L. Princeton. *W. W. B. & J. F. C.*
Muhlenbergii, Schkuhr. Open sterile soil.
cephaloidea, Dewey. Wet meadows.
cephalophora, Muhl. Dry soil.
echinata, Murray, var. cephalantha, Bailey. Rare.
" " var. *microstachys, Bœckl.
Princeton. *W. W. B. & J. F. C.*
canescens, L. Swamps.
" " var. vulgaris, Bailey. Common.
trisperma, Dewey. Swamps.
Deweyana, Schwein. Worcester. *Prof. Eaton.*
bromoides, Schkuhr. Swampy woods.

tribuloides, Wahl., var. cristata, Bailey.   Worcester.
                                              *Prof. Eaton.*
scoparia, Schkuhr.                              Meadows.
straminea, Willd.           Dry fields.   *Prof. Eaton.*
"          "    var. mirabilis, Tuckerm.
                     Princeton.   *W. W. B. & J. F. C.*

## GRAMINEÆ.

PASPALUM, L.   Paspalum.
  setaceum, Michx.                              Pastures.
  læve, Michx.                                *G. E. Stone.*
PANICUM, L.   Panic-Grass.
  *glabrum, Gaudin.                             Pastures.
  *sanguinale, L.   Common Crab-Grass.          Common.
  capillare, L.   Old-witch Grass.                "
  agrostoides, Muhl.              Sandy margins of ponds.
  xanthophysum, Gray.                           Princeton.
  latifolium, L.                             Moist thickets.
  depauperatum, Muhl.                           Dry hills.
  dichotomum, L.                                Common.
  *Crus-galli, L.   Barnyard-Grass.               "
  *miliaceum, L.                             East Worcester.
SETARIA, Beauv.   Bristly Foxtail Grass.
  *glauca, Beauv.   Foxtail.                    Not rare.
  *viridis, Beauv.   Green Foxtail.              "      "
  *Italica, Kunth.   Hungarian Grass.   Rarely spontaneous.
CENCHRUS, L.   Bur-Grass.
  tribuloides, L.                              Sandy soil.
LEERSIA, Swartz.   White Grass.
  oryzoides, Swartz.   Rice Cut-grass.          Swamps.

ANDROPOGON, Royen.   Beard-Grass.
  furcatus, Muhl.                       Dry soil. Not rare.
  scoparius, Michx.                     Poor soil. Common.
  Virginicus, L.                        Princeton. *Prof. Eaton.*
CHRYSOPOGON, Trin.   Broom-Corn.
  nutans, Benth.  Wood-Grass.           Quite common.
PHALARIS, L.   Canary-Grass.
  *Canariensis, L.  Canary-Grass.       Waste places.
  arundinacea, L.  Reed Canary-Grass.   Wet grounds.
ANTHOXANTHUM, L.   Sweet Vernal-Grass.
  *odoratum, L.                         Meadows, common.
STIPA, L.   Feather-Grass.
  Richardsonii, Link.   Mt. Wachusett. *Prof. Eaton.*
ORYZOPSIS, Michx.   Mountain Rice.
  melanocarpa, Muhl.   Princeton. *W. W. B. & J. F. C.*
  asperifolia, Michx.                   Mt. Wachusett.
  Canadensis, Torr.    West Boylston. *Prof. Eaton.*
MUHLENBERGIA, Schreber.   Drop-seed Grass.
  glomerata, Trin.     Southbridge. *L. E. Ammidown.*
  Mexicana, Trin.      Princeton. *W. W. B. & J. F. C.*
  sylvatica, Torr. & Gray.              Copses, common.
  Willdenovii, Trin.                    "   not rare.
  diffusa, Schreber.  Drop-seed.        Dry hills.
BRACHYELYTRUM, Beauv.
  aristatum, Beauv.                     Copses.
PHLEUM, L.   Cat's-tail Grass.
  *pratense, L.  Timothy.               Common.
ALOPECURUS, L.   Foxtail Grass.
  *pratensis, L.  Meadow Foxtail.       Common.

geniculatus, L., var. aristulatus, Torr.  In very wet places.
SPOROBOLUS, R. Br.    Drop-seed Grass.
  vaginæflorus, Vasey.                              Dry fields.
AGROSTIS, L.    Bent-Grass.
  *alba, L.    White Bent-Grass.                    Common.
  * "       "   var. vulgaris, Thurber.  Red Top.   "
  perennans, Tuckerm.    Thin-Grass.                Damp places.
  scabra, Willd.    Hair-Grass.                     Dry places.
CINNA, L.    Wood Reed-Grass.
  arundinacea, L.                                   Woods, not rare.
  pendula, Trin.      Princeton.   *W. W. B. & J. F. C.*
CALAMAGROSTIS, Adans.    Reed Bent-Grass.
  Canadensis, Beauv.    Blue-Joint Grass.           Meadows.
  Nuttalliana, Steud.                               Moist grounds.
ARRHENATHERUM, Beauv.    Oat-Grass.
  *avenaceum, Beauv.                                Meadows.
HOLCUS, L.    Meadow Soft-Grass.
  *lanatus, L.    Velvet-Grass.                     Moist meadows.
DESCHAMPSIA, Beauv.    Hair-Grass.
  flexuosa, Trin.    Common Hair-Grass.             Mt. Wachusett.
AVENA, Tourn.    Oat.
  *sativa, L.                                       Occasionally spontaneous.
  striata, Michx.                                   *G. E. Stone.*
DANTHONIA, DC.    Wild Oat-Grass.
  spicata, Beauv.                                   Pastures.
PHRAGMITES, Trin.    Reed.
  communis, Trin.      Westborough.   *A. N. Randlett.*
ERAGROSTIS, Beauv.
  *major, Host.                                     Waste places.

Purshii, Schrader.     *G. E. Stone.*
pectinacea, Gray.     Pastures.
DACTYLIS, L.     Orchard Grass.
    *glomerata, L.     Common.
BRIZA, L.     Quaking Grass.
    *media, L.     Pastures.
POA, L.     Meadow-Grass.
    *annua, L.     Low Spear-Grass.     Common.
    serotina, Ehrh.     False Red-top.     "
    pratensis, L.     June Grass.     "
GLYCERIA, R. Br.     Manna-Grass.
    Canadensis, Trin.     Rattlesnake-Grass.     Wet places.
    obtusa, Trin.     "    "
    nervata, Trin.     Fowl Meadow-Grass.     Moist meadows.
    fluitans, R. Br.     Shallow water.
    acutiflora, Torr.     Wet places.
FESTUCA, L.     Fescue-Grass.
    ovina, L.     Sheep's Fescue.     *G. E. Stone.*
    nutans, Willd.     Princeton.     *W. W. B. & J. F. C.*
    *elatior, L.     Meadow Fescue.     Rich grassland.
    "    " var. pratensis, Gray.     Princeton.
                                               *W. W. B. & J. F. C.*
BROMUS, L.     Brome-Grass.
    *secalinus, L.     Cheat.     Waste grounds.
    *racemosus, L.     Upright Chess.     "    "
    ciliatus, L.     River banks, etc.
    *asper, L.     Princeton.     *W. W. B. & J. F. C.*
    *tectorum, L.     Waste grounds.
LOLIUM, L.     Darnel.
    perenne, L.     Common Darnel.     Fields, etc.

AGROPYRUM, Gærtn.   Wheat-Grass.
  repens, Beauv.   Quitch-Grass.           Common
ELYMUS, L.   Wild Rye.
  Virginicus, L.                            Wet places
  Canadensis, L.                            "      "
ASPRELLA, Willd.   Bottle-brush Grass.
  Hystrix, Willd.                           Moist woodlands

# SERIES II. CRYPTOGAMIA.
## CLASS III. ACROGENÆ.
### SUBCLASS I. PTERIDOPHYTA.

#### EQUISETACEÆ.

EQUISETUM, L.   Horsetail.
  arvense, L.   Common Horsetail.          Gravelly soil.
  sylvaticum, L.                           Moist shaded places.
  limosum, L.                              In shallow water.
  hyemale, L.                              *G. E. Stone.*
  scirpoides, Michx.   Southbridge.  *L. E. Ammidown.*

#### FILICES.

POLYPODIUM, L.   Polypody.
  vulgare, L.                              Rocks, common.
ADIANTUM, L.   Maidenhair.
  pedatum, L.                              Rich moist woods.
PTERIS, L.   Brake.
  aquilina, L.   Common Brake.             Thickets, common.
ASPLENIUM, L.   Spleenwort.
  Trichomanes, L.                          Rare.

ebeneum, Ait.                                    Common.
thelypteroides, Michx.                        "
Filix-fœmina, Bernh.                          "

PHEGOPTERIS, Fée.     Beech Fern.
polypodioides, Fée.                           Damp woods.
hexagonoptera, Fée.                           "      "
Dryopteris, Fée.                                 "      "

ASPIDIUM, Swartz.     Shield Fern.
Thelypteris, Swartz.                          Woodlands, common.
Noveboracense, Swartz.                     "      "
spinulosum, Swartz, var. intermedium, D. C. Eaton.
                                                               Common.
cristatum, Swartz.                         Quite    "
    "      "    var. Clintonianum.        Worcester.
Goldianum, Hook.     Spencer.    *Miss A. E. Tucker.*
marginale, Swartz.                Rich woods, common.
acrostichoides, Swartz.   Christmas Fern.       "

CYSTOPTERIS, Bernh.     Bladder Fern.
fragilis, Bernh.                                 Worcester.

ONOCLEA, L.     Onoclea.
sensibilis, L.   Sensitive Fern.         Very common.
    "      "   var. *obtusilobata.   Princeton.
                                                             *W. W. B. & J. F. C.*
Struthiopteris, Hoffm.   Ostrich Fern.       Worcester.

WOODSIA, R. Br.     Woodsia.
Ilvensis, R. Br.          Spencer.    *Miss A. E. Tucker.*
obtusa, Torr.                                 *Miss Wheelock.*

DICKSONIA, L'Her.     Dicksonia.
pilosiuscula, Willd.       Moist shady places, common.

LYGODIUM, Swartz.   Climbing Fern.
palmatum, Swartz.           Uxbridge, Oxford.   Rare.
OSMUNDA, L.   Flowering Fern.
regalis, L.                       Swamps, common.
Claytoniana, L.               Low grounds,   "
cinnamomea, L.   Cinnamon Fern.            "

### OPHIOGLOSSACEÆ.

BOTRYCHIUM, Swartz.   Moonwort.
lanceolatum, Angstrœm. Princeton. *W. W. B. & J. F. C.*
matricariæfolium, Braun.   "        "      "    "
ternatum, Swartz.                        Pastures.
"         "   var. lunarioides.           "
"         "    "  intermedium.            "
"         "    "  dissectum.              "
Virginianum, Swartz.                 Rich woods.
OPHIOGLOSSUM, L.   Adder's Tongue.
vulgatum, L.   Worcester.   Spencer, *Miss A. E. Tucker.*

### LYCOPODIACEÆ.

LYCOPODIUM, L.   Club-Moss.
Selago, L.                              Mt. Watatic.
lucidulum, Michx.   Staghorn Moss.        Common.
inundatum, L.                          Sandy shores.
annotinum, L.                Leicester.   *G. Coult.*
obscurum, L.                 Moist woods, common.
clavatum, L.   Common Club-Moss.          "
complanatum, L.   Ground Pine.            "
         "   " var. *Chamæcyparissus.   Leicester.
                                          *G. Coult.*

## SELAGINELLACEÆ.

SELAGINELLA, Beauv.    Dwarf Club-Moss.
   rupestris, Spring.        Worcester.    *G. T. Rignel.*
   apus, Spring.             Southbridge.   *L. E. Ammidown.*
ISOETES, L.    Quillwort.
   lacustris, L.                              *G. E. Stone.*
   echinospora, Durieu, var. Braunii, Engelm.   " " "
   riparia, Engelm.                             " " "


## NOTE.

THE names of the following species were either accidentally omitted or received too late to be inserted in the proper places indicated below:

PAGE 264.

HYPERICUM, Tourn.
   Canadense, L., var. *majus, Gray.   Princeton.
                                       *W. W. B. & J. F. C.*
ABUTILON, Tourn.    Indian Mallow.
   *Avicennæ, Gærtn.   Velvet-Leaf.    Escaped.
                          Southbridge.   *L. E. Ammidown.*
HIBISCUS, L.    Rose Mallow.
   *Trionum, L.   Bladder Ketmia.   Southbridge.
                                         *L. E. Ammidown.*

PAGE 266.

VITIS, Tourn.
   æstivalis, Michx.   Summer Grape.              Common.

PAGE 269.

LATHYRUS, Tourn.
 palustris, L.  Berlin. *Miss J. M. Nichols.*

PAGE 273.

CALLITRICHE, L.
 heterophylla, Pursh. Mt. Wachusett. *W. W. B. & J. F. C.*

PAGE 286.

RHODODENDRON, L.
 viscosum, Torr., var. *glaucum, Gray.
  Princeton. *W. W. B. & J. F. C.*

PAGE 299.

QUERCUS, L.
 palustris, DuRoi. Pin Oak.  In the southern part of the county.

The hope expressed on page 285 concerning *Andromeda polifolia*, L. has been fulfilled since that page was printed. October 18, 1894, Mr. G. S. Newcomb of Westborough reported to me the finding of this shrub in a swamp close by that village, and a few days later I had the pleasure of visiting the locality and of finding it in abundance.

*\** I shall be glad to receive any information as to the discovery of species hitherto unrecorded within our limits, or any facts concerning the distribution of our rarer species already recorded.

The preceding list contains 1,098 species and varieties, of which 55 are cryptogams.

# APPENDIX B.

The trees, shrubs and evergreen flowering plants growing without cultivation in Worcester County.

| | |
|---|---|
| Coptis trifolia, Salisb. | Goldthread. |
| Liriodendron Tulipifera, L. | Tulip-Tree. |
| Berberis vulgaris, L. | Barberry. |
| Tilia Americana, L. | Basswood. |
| Xanthoxylum Americanum, Mill. | Northern Prickly Ash. |
| Ilex verticillata, Gray. | Black Alder. |
| lævigata, Gray. | Smooth Winterberry. |
| Nemopanthes fascicularis, Raf. | Mountain Holly. |
| Celastrus scandens, L. | Climbing Bitter-sweet. |
| Rhamnus cathartica, L. | Common Buckthorn. |
| Ceanothus Americanus, L. | New Jersey Tea. |
| Vitis Labrusca, L. | Northern Fox-Grape. |
| æstivalis, Michx. | Summer Grape. |
| cordifolia, Michx. | Frost Grape. |
| riparia, Michx. | River Grape. |
| Ampelopsis quinquefolia, Michx. | Virginian Creeper. |
| Acer Pennsylvanicum, L. | Striped Maple. |
| spicatum, Lam. | Mountain " |
| saccharinum, Wang. | Rock " |

| | |
|---|---|
| dasycarpum, Ehrh. | White Maple |
| rubrum, L. | Red " |
| Rhus typhina, L. | Staghorn Sumach |
| glabra, L. | Smooth " |
| copallina, L. | Dwarf " |
| venenata, DC. | Poison Dogwood |
| Toxicodendron, L. | Poison Ivy |
| "         " var. radicans, L. | "         " |
| Genista tinctoria, L. | Dyer's Greenweed |
| Robinia Pseudacacia, L. | Common Locust |
| viscosa, Vent. | Clammy " |
| Prunus pumila, L. | Dwarf Cherry |
| Pennsylvanica, L. f. | Wild Red " |
| Virginiana, L. | Choke " |
| serotina, Ehrh. | Wild Black " |
| Spiræa salicifolia, L. | Common Meadow-Sweet |
| tomentosa, L. | Hardhack |
| Physocarpus opulifolius, Maxim. | Nine-bark |
| Rubus odoratus, L. | Purple Flowering-Raspberry |
| strigosus, Michx. | Wild Red " |
| occidentalis, L. | Thimbleberry |
| villosus, Ait. | High Blackberry |
| Canadensis, L. | Low " |
| hispidus, L. | Swamp " |
| Potentilla fruticosa, L. | Shrubby Cinquefoil |
| Rosa blanda, Ait. | Wild Rose |
| Carolina, L. | "         " |
| lucida, Ehrh. | "         " |
| humilis, Marsh. | "         " |

## LIST OF TREES AND SHRUBS. 323

    rubiginosa, L.                                       Sweetbrier.
Pyrus arbutifolia, L. f.                       Choke-berry.
    "    " " var. melanocarpa, Hook.    "
    Americana, DC.           American Mountain-Ash.
    aucuparia, Gærtn.           European   "
    Malus, L.                                  Wild Apple.
    communis, L.                       " Pear.
Cratægus coccinea, L.                White Thorn.
    "   " var. mollis, Torr. & Gray.   "  "
    punctata, Jacq.                      "  "
Amelanchier Canadensis, Torr. & Gray.     Shad-bush.
Hydrangea arborescens, L.        Wild Hydrangea.
Ribes Cynosbati, L.                Gooseberry.
    rotundifolium, Michx.              "
    oxyacanthoides, L.                 "
    prostratum, L'Her.             Fetid Currant.
    floridum, L'Her.              Wild Black   "
    rubrum, L., var. subglandulosum, Maxim.   Red   "
Hamamelis Virginiana, L.              Witch-Hazel.
Decodon verticillatus, Ell.          Swamp Loosestrife.
Aralia hispida, Vent.                Bristly Sarsaparilla.
Cornus florida, L.                  Flowering Dogwood.
    circinata, L'Her.              Round-leaved Cornel.
    sericea, L.                      Silky   "
    stolonifera, Michx.            Red-osier Dogwood.
    paniculata, L'Her.             Panicled Cornel.
    alternifolia, L. f.              Alternate-leaved   "
Nyssa sylvatica, Marsh.                Tupelo.

| | |
|---|---|
| Sambucus Canadensis, L. | Common Elder. |
| ′ racemosa, L. | Red-berried " |
| ′ Viburnum lantanoides, Michx. | Hobble-bush. |
| ' Opulus, L. | Cranberry-tree. |
| ( acerifolium, L. | Arrow-wood. |
| . dentatum, L. | " |
| ' cassinoides, L. | Withe-rod. |
| ✳ Lentago, L. | Sweet Viburnum. |
| - Linnæa borealis, Gronov. | Twin-flower. |
| ′ Symphoricarpus racemosus, Michx. | Snowberry. |
| ′ Lonicera ciliata, Muhl. | Fly-Honeysuckle. |
| ( cærulea, L. | Mountain " " |
| hirsuta, Eaton. | Hairy " |
| ' glauca, Hill. | " |
| ✳ Diervilla trifida, Mœnch. | Bush Honeysuckle. |
| ✳ Cephalanthus occidentalis, L. | Button-bush. |
| ✳ Mitchella repens, L. | Partridge-berry. |
| ′ Gaylussacia dumosa, Torr. & Gray. | Dwarf Huckleberry. |
| ′ frondosa, Torr. & Gray. | Dangleberry. |
| ′ resinosa, Torr. & Gray. | Huckleberry. |
| ′ Vaccinium Pennsylvanicum, Lam. | Dwarf Blueberry. |
| ( Canadense, Kalm. | Canada " |
| vacillans, Solander. | Low " |
| ·. corymbosum, L. & vars. | High " |
| ' Oxycoccus, L. | Small Cranberry. |
| . macrocarpon, Ait. | Large " |
| ′ Chiogenes serpyllifolia, Salisb. | Creeping Snowberry. |
| ( Arctostaphylos Uva-ursi, Spreng. | Bearberry. |
| ′ Epigæa repens, L. | Trailing Arbutus. |
| ′ Gaultheria procumbens, L. | Checkerberry. |

## LIST OF TREES AND SHRUBS. 325

| | |
|---|---|
| Andromeda polifolia, L. | Water Andromeda. |
| ligustrina, Muhl. | Andromeda. |
| Leucothoë racemosa, Gray. | Leucothoë. |
| Cassandra calyculata, Don. | Leather-Leaf. |
| Kalmia latifolia, L. | Mountain Laurel. |
| angustifolia, L. | Sheep " |
| glauca, Ait. | Pale " |
| Rhododendron viscosum, Torr. | White Azalea. |
| " " var. glaucum, Gray. | White Azalea. |
| nudiflorum, Torr. | Swamp Pink. |
| Rhodora, Don. | Rhodora. |
| maximum, L. | Rhododendron. |
| Ledum latifolium, Ait. | Labrador Tea. |
| Clethra alnifolia, L. | Sweet Pepperbush. |
| Chimaphila umbellata, Nutt. | Prince's Pine. |
| maculata, Pursh. | Spotted Wintergreen. |
| Moneses grandiflora, Salisb. | One-flowered Pyrola. |
| Pyrola secunda, L. | Wintergreen. |
| chlorantha, Swartz. | " |
| elliptica, Nutt. | Shin-leaf. |
| rotundifolia, L. | Wintergreen. |
| Fraxinus Americana, L. | White Ash. |
| sambucifolia, Lam. | Black Ash. |
| Sassafras officinale, Nees. | Sassafras. |
| Lindera Benzoin, Blume. | Spice-bush. |
| Dirca palustris, L. | Leatherwood. |
| Daphne Mezereum, L. | Mezereum. |
| Ulmus fulva, Michx. | Slippery Elm. |
| Americana, L. | American " |

' Celtis occidentalis, L.     Hackberry.
    Mr. G. A. Cheney assures me that he has found it growing in Sturbridge and elsewhere in the county.

| | |
|---|---|
| Morus alba, L. | White Mulberry. |
| Platanus occidentalis, L. | Buttonwood. |
| ' Juglans cinerea, L. | Butternut. |
| ' Carya alba, Nutt. | Shag-bark Hickory. |
| '    porcina, Nutt. | Pig-nut " |
| '    amara, Nutt. | Bitter-nut " |
| · Myrica Gale, L. | Sweet Gale. |
| '    cerifera, L. | Bayberry. |
| '    asplenifolia, Endl. | Sweet Fern. |
| ' Betula lenta, L. | Black Birch. |
| '    lutea, Michx. f. | Yellow " |
| '    populifolia, Ait. | Gray " |
| '    papyrifera, Marshall. | Paper " |
| '    nigra, L. | Red " |
| ' Alnus incana, Willd. | Speckled Alder. |
|     serrulata, Willd. | Smooth " |
| ' Corylus Americana, Walt. | Wild Hazel-nut. |
| '    rostrata, Ait. | Beaked " |
| ' Ostrya Virginica, Willd. | Hop-Hornbeam. |
| ' Carpinus Caroliniana, Walt. | Hornbeam. |
| ' Quercus alba, L. | White Oak. |
|     macrocarpa, Michx. | Bur " |

    I am indebted to Mrs. H. G. Waite for the addition of this species.

| | |
|---|---|
| '    bicolor, Willd. | Swamp White Oak. |
| ·    Prinus, L. | Chestnut " |
|     "   " var. monticola, Michx. | Rock "   " |

# LIST OF TREES AND SHRUBS.

- prinoides, Willd.            Dwarf Chestnut Oak.
- rubra, L.            Red "
- coccinea, Wang.            Scarlet "
- "    " var. tinctoria, Gray.      Black "

    palustris, DuRoi.            Pin "

      Miss A. H. Tucker in *Trees of Worcester*.

- ilicifolia, Wang.            Scrub "

Castanea sativa, Mill., var. Americana, Michx.     Chestnut.

Fagus ferruginea, Ait.            Beech.

Salix nigra, Marsh.            Black Willow.

    lucida, Muhl.            Shining "

    fragilis, L.            Crack "

    alba, L.            White "

    "    " var. vitellina, Koch.      " "

    rostrata, Richardson.            Beaked "

    discolor, Muhl.            Glaucous "

- humilis, Marsh.            Prairie "
- tristis, Ait.            Dwarf Gray "

    sericea, Marsh.            Silky "

- cordata, Muhl.            Heart-leaved "

    myrtilloides, L.            "

Populus tremuloides, Michx.            American Aspen.

    grandidentata, Michx.            Large-toothed "

    balsamifera, L., var. candicans, Gray.     Balm of Gilead.

    monilifera, Ait.            Cottonwood.

Pinus Strobus, L.            White Pine.

    rigida, Mill.            Pitch "

    resinosa, Ait.            Red "

Picea nigra, Link.            Black Spruce.

## FLORA OF WORCESTER COUNTY.

| | | |
|---|---|---|
| ' | Tsuga Canadensis, Carr. | Hemlock. |
| ' | Abies balsamea, Mill. | Balsam Fir. |
| ' | Larix Americana, Michx. | Hackmatack. |
| ' | Chamæcyparis sphæroidea, Spach. | White Cedar. |
| ' | Juniperus communis, L. | Common Juniper. |
| |     Sabina, L., var. procumbens, Pursh. | " |
| ' |     Virginiana, L. | Red Cedar. |
| ' | Taxus Canadensis, Willd. | Ground Hemlock. |
| ' | Smilax rotundifolia, L. | Common Greenbrier. |

# INDEX.

Achillea, 169.
Actæa, 113.
Adiantum, 102.
Agrostis, 77.
Alopecurus, 42.
Amelanchier, 37.
Ampelopsis, 85, 213.
Amphicarpæa, 101, 237.
Andromeda, 39.
Andropogon, 114.
Anemone, 23, 30.
Anemophilous flowers, 29.
Anthoxanthum, 47.
Antrostomus, 44.
Apios, 101.
Apocynum, 191.
Aralia, 112.
Arbutus, Trailing, 30.
Arctostaphylos, 39.
Arethusa, 49, 65.
Arisæma, 44.
Arnold, Matthew, 89, 91, 137.
Arrhenatherum, 80.
Ash, Northern Prickly, 170.
Aspidium, 201.
Asplenium, 102, 238.
Asters, 98, 114.

Barbauld, Mrs., 11.
Basswood, 223.
Bearberry, 39.
Bidens, 249.
Birches, 155.
Bittersweet, 171.
Bloodroot, 30.

Blueberry, 38.
Bobolink, 58, 59.
Bonasa, 44.
Botrychium, 67.
Brachyelytrum, 80.
Branchipus, 33.
Britten & Holland, 136.
Briza, 79, 183.
Brown, Robert, 48, 86.
Bryant, 106, 125, 131, 158, 175.
Buda, 68.
Bunyan, 136.
Burns, 17.
Burroughs, 169, 185.
Button-bush, 72, 224.

Capsella, 50.
Cardinal-flower, 102.
Carex, 203.
Carpinus, 145.
Carpodacus, 58.
Cassandra, 23.
Cassia Braziliana, 109.
Castanea, 84.
Castilleia, 69.
Catbird, 56.
Caulophyllum, 48.
Cenchrus, 237.
Cephalanthus, 72.
Chaucer, 16.
Chelone, 101.
Chewink, 44.
Chiogenes, 39.
Chrysanthemum, 68.
Chrysomitris, 58.

Chrysopogon, 114, 247.
Cinna, 115, 247.
Circæa, 203.
Claytonia, 47.
Clethra, 85, 235.
Clintonia, 157.
Cnicus, 84.
Cocos, 27.
Colias, 32.
Collier, W. F., 2.
Corallorhiza, 100.
Cornus, 64.
Corydalis, 162.
Corylus, 24, 125.
Coues, Elliott, 44, 56.
Cowper, 10.
Cuscuta, 100.
Cyanospiza, 58.
Cynoglossum, 44.
Cynthia, 83.
Cyperus, 81, 82, 247.
Cypripedium, 66.
Cystopteris, 249.

Dactylis, 77.
Daisy, Ox-eye, 68.
Dandelion, 49.
Darwin, 47, 49, 147.
Desmodium, 99.
Dirca, 31.
Dolichonyx, 58.
Dragon-flies, 82.
Drosera, 49, 87.

Earle, J., 205.
Eaton, D. C., 200.
Emerson, G. B., 27.
Emerson, R. W., 18, 19, 30, 51, 61, 110, 191, 227.
Entomophilous flowers, 29.
Epigæa, 30.
Equisetum, 145.
Erigeron, 42.
Eriocaulon, 86.
Eriophorum, 158.
Erythronium, 122.
Eupatorium, 98.
Euphorbia, 226.
Ewing, Mrs. J. H., 35.

Ferns, 200.

Flagg, Wilson, 37, 53, 59, 191, 224
Floras of N. A., 208–210.

Galinsoga, 250.
Gaultheria, 63.
Gaylussacia, 37, 38, 110.
Gentiana, 102, 115.
Geothlypis, 58.
Gerardia, 99, 227.
Gilpin, William, 95.
Gœthe, 149.
Golden-rods, 97.
Goodyera, 100, 236.
Grapta, 32.
Grasses, 76.
Gratiola, 102.
Gray, Asa, 47, 148.
Grosbeak, 161.
Grouse, 44.

Habenaria, 49, 65, 66, 86, 99.
Hamamelis, 24.
Harporhynchus, 57.
Hazel, 24, 125.
Heliopsis, 250.
Hepatica, 30.
Hierochloe, 75.
Higginson, T. W., 102, 214, 217.
Hipparchia, 84.
Holcus, 80.
Holmes, O. W., 218.
Homer, 9, 17.
Honeysuckle, 156.
Hooker, Sir J. D., 131.
Hooker, Sir W. J., 48.
Hooker & Baker, 200.
Huckleberry, 37.
Hypoxis, 157.

Icones Carpologicæ, 109.
Icterus, 58.
Ilex, 112.
Ivy, poison, 171.

Jackson, Helen, 115, 146, 240.
Jefferies, R., 121, 123, 187.

Kalmia, 21, 62.
Keats, 104.
King, T. Starr, 231.
Kingsley, Charles, 7, 40, 172, 241.

## INDEX.

Labrador Tea, 26.
Landor, W. S., 169.
Laurel, 21, 62.
Ledum, 26.
Leersia, 247.
Lespedeza, 99.
Leucothoe, 42.
Lichens, 126.
Lilium, 86.
Linnæa, 60.
Linnæus, 40, 60, 208.
Liparis, 65.
Lobelia, 102.
Longfellow, 155, 156.
Ludwigia, 114.
Lycæna, 32.
Lygodium, 249.

Maple, Striped, 154.
Marchantia, 27.
Massee, G., 139.
Meadow-sweet, 193.
Medicago, 167.
Merula, 53.
Michaux, 208.
Microstylis, 213.
Milton, 133, 135, 136.
Mimus, 56.
Moneses, 27.
Monotropa, 100.
Moore, 66.
Morris, William, 25, 41, 193.
Muhlenbergia, 114.

Nasturtium, 22.
Nymphalis, 84.

Oakesia, 123.
Oriole, 58.
Osmunda, 66.
Ostrya, 145.
Ovenbird, 44, 168.

Packard, A. S., 33.
Painted-cup, 69.
Panicum, 114.
Parkman, F., 180.
Parnassia, 101, 250.
Phleum, 77.
Phytolacca, 112.
Pine, White, 191.

Pipilo, 43.
Pitcher-plant, 180.
Poa, 77, 132.
Pogonia, 45, 49, 65.
Polygala, 23.
Pyranga, 44.
Pyrola, 202.
Pyrus, 112.

Quercus, 22.

Ranunculus, 42, 220.
Rhexia, 86.
Rhododendron, 69, 224.
Rhus, 85, 171.
Ribes, 145.
Robin, 53.
Rosa, 60.
Rosaceæ, 192, 193.
Rubus, 224.
Rudbeckia, 68.
Rumex, 50.
Ruskin, 127.

Sagittaria, 237.
Sambucus, 110, 145.
Sanguinaria, 30.
Sarracenia, 180.
Sarrazin, Dr., 180.
Sassafras, 154.
Scirpus, 204.
Scudder, S. H., 83.
Sedges, 80.
Seiurus, 44.
Setaria, 114.
Setophaga, 58.
Shakspere, 134, 101.
Shepherd's Purse, 132.
Sialia, 58.
Sium, 248.
Smilax, 161.
Snowberry, Creeping, 39.
Solanum, 171.
Solidago, 97.
Sophocles, 91.
Sparrow, Chipping, 55.
Spergularia, 68.
Spiranthes, 99, 250.
Spizella, 55.
Stellaria, 50.

Sumach, 85.
Sundew, 87.
Symplocarpus, 23.

Tanager, Scarlet, 44, 45.
Taraxacum, 49.
Tennyson, 13, 28, 71, 81, 83, 88, 141, 193, 242.
Tetraphis, 31.
Theocritus, 105.
Thomson, J., 195.
Thoreau, 160, 161, 184, 187.
Thornton, 134.
Thrasher, Brown, 57.
Tissa, 68.
Trientalis, 23.
Trillium, 35, 36.
Trolls, 7, 8.

Tupelo, 179.
Typha, 13.

Utricularia, 215.

Vaccinium, 38, 111.
Vanessa, 32, 83, 84.
Viburnum, 36, 65, 111.
Viola, 30, 134–140.

Walton, Izaak, 94.
Water-lilies, 214.
Whippoorwill, 45, 46.
White, Gilbert, 185.
Whittier, J. G., 228.
Witch-hazel, 116.
Wordsworth, 5, 16, 32, 115, 193.

Xyris, 248.

www.ingramcontent.com/pod-product-compliance
Lightning Source LLC
Chambersburg PA
CBHW020312240426
43673CB00039B/780